Peter Guthrie Tait

Lectures on Some Recent Advances in Physical Science

With a special Lecture on Force. Third Edition

Peter Guthrie Tait

Lectures on Some Recent Advances in Physical Science
With a special Lecture on Force. Third Edition

ISBN/EAN: 9783337776022

Printed in Europe, USA, Canada, Australia, Japan

Cover: Foto ©berggeist007 / pixelio.de

More available books at **www.hansebooks.com**

LECTURES ON PHYSICAL SCIENCE

LECTURES

ON SOME

RECENT ADVANCES

IN

PHYSICAL SCIENCE

WITH A SPECIAL LECTURE ON

FORCE

BY

P. G. TAIT, M.A.

FORMERLY FELLOW OF ST. PETER'S COLLEGE, CAMBRIDGE
PROFESSOR OF NATURAL PHILOSOPHY IN THE UNIVERSITY OF EDINBURGH

THIRD EDITION. REVISED

London
MACMILLAN AND CO
1885

WITH THIS WORK

I DESIRE TO ASSOCIATE THE NAMES

OF

GEORGE BARCLAY AND THOMAS STEVENSON,

FELLOWS OF THE ROYAL SOCIETY OF EDINBURGH,

BY WHOSE EFFORTS THESE LECTURES WERE ORGANISED,

AND AT WHOSE WISH THEY ARE PUBLISHED AS DELIVERED.

P. G. T.

PREFACE TO THIRD EDITION.

IN preparing a Third Edition for the press, I have adhered to my original plan of publishing these Lectures just as they were taken down by the short-hand writers. I have, however, altered here and there a mere word or two, and in a few places, where it appeared to be called for, I have added an explanatory sentence.

Other brief additions [enclosed in square brackets] deal chiefly with facts which have been discovered since the Second Edition (a very large one) was published.

I have not reprinted the polemical part of the Preface to that edition. Professor Zöllner's charges, there alluded to, were withdrawn by himself:—while those of Professor Clausius were so fully met by me in the *Philosophical Magazine* for May 1879 that his reply has not, so far as I know, even yet appeared. And the reference to Mohr's work has been amplified, and embodied in the text of the book.

Here my Preface might have ended, had it not been that a new critic has appeared on the scene, in the form of Professor du Bois-Reymond, who, in his capacity of Secretary to the *Royal Academy of Sciences of Berlin*, considered himself justified in speaking as follows at a gala meeting of that Academy on March 28th, 1878 :[1]—

'Foreign investigators, in their ignorance of the German language, often discovered for the second time things long known to us.

'Not unfrequently, even when better informed, they took advantage of the presumed right of independent discovery to cite their German predecessors only by the way or not at all. The Germans, on the other hand, showed a perfect national impartiality which was far more to their credit than their linguistic superiority. Indeed they never even conceived the possibility of national jealousy between learned men who seek nothing but the One Truth, but live, ideally, with the investigators of all countries as with their equals, without even imagining how little this feeling is reciprocated, chiefly because foreigners know so little of us.

'In other nations great pains were taken to find out among themselves the germs of new discoveries, and in one way or another this always succeeded. The German man of science only wished to find the true germ, whether it might be in a fellow-countryman or in a foreigner, and he never hesitated to recognise, as probably the first discoverer, a foreigner, if there was the slightest reason for the supposition. He was far more pleased to do historical justice than hurt to deprive Germany of a doubtful glory.

'In the same way it was far from the thought of German men of science to exaggerate the importance of a first chance observation, in order out of it to add to Germany's scientific credit.

'What weight would others not have given to the fact, quite unnoticed by us, that the first galvanic phenomenon, which besides gave Volta the key to Galvani's researches, was observed here in Berlin by one of our predecessors?

'The national feeling does not blind German scientific men to the fact that the seeking out of such Priority is a double-edged weapon. For if an Irish physicist living in England and a Scottish physicist (who need no such addition to their fame) had Spectrum Analysis in their pocket ten years before Kirchhoff and

---

[1] The obviously offensive intention which dictated this speech rendered me anxious to avoid all suspicion of having heightened it in translation :— so, at my request, my colleague Dr. Crum Brown has kindly made the subjoined version for me.

Bunsen, why did they not make out of it what Bunsen and Kirchhoff did?
'Why? A Scottish man of science, whose name has been recently much before us, tells us in his *Lectures on some Recent Advances in Physical Science.* The German investigator knows all that is going on in Science, or at least has some one by him who does. If a German comes on a new idea, he can at once see, or be told, whether another has it or not, and in the latter case he can print the idea, and so secure the priority: the poor Britons, on the other hand, make the most splendid discoveries in the world without ever guessing that they have struck on anything new—like the *Bourgeois Gentilhomme*, they speak prose without knowing it,—and let the priority slip them. The wily Germans! who, instead of contenting themselves like other innocent folks with their mother tongue, sneak into foreign languages to spy out the discoveries that are being made.

'The unpleasant impression produced by these statements in the key of national antipathy is increased by other passages in these *Lectures*. The author makes it his special business to elucidate the history of the law of the conservation of energy, and tracks this law back to Newton's third law of motion,—the equality of action and reaction. Newton's second explanation of his third law is, he tells us, a nearly complete expression of the conservation of energy.

'As the science of Mechanics depends on Newton's laws of motion, of course the conservation of energy can be somehow read out of them, or rather read into them. And we need not doubt that a head like Newton's had, in private, as much knowledge of the conservation of energy as could be had in his time. It is another question what view he took of it, and what was his position towards it as manifested in his works. Whoever is acquainted with the history of this doctrine knows Descartes's original but unsuccessful notions; their correction by Leibniz: Leibniz's conception of the material world substantially agreeing with that now held. He knows that Newton in his Optics also disproved Descartes's opinion, although without mentioning its correction by Leibniz, and without himself undertaking this correction; that the Cosmogony-speculator called in God to put the planetary system right when it had gone wrong in consequence of accumulated perturbations, which scarcely accords with the conservation of energy. To one who knows this epoch it will not seem impossible that the dissensions between Leibniz and Newton disgusted the latter with the subject, and formed the cause why the law of the conservation of energy received then so little assent in England. Certain it is that on the Continent, during the first half of last century, this law in the form given to it by Leibniz was the common property of scientifically educated persons, as it is now. This is no hidden mystery: it is easy to make it out from the literature of the last ten years. He who has all this before him can only shrug his shoulders at the artificial attempts to put Newton at the head of those to whom we owe the law of the conservation of energy. Perhaps the author of the *Lectures* is not sufficiently acquainted with the history on which he undertakes to throw light, and on the later developments of which he passes such rough judgment, and so only lays himself open to the suspicion, unfortunately not weakened by his other writings, that the fiery Celtic blood of his country sometimes runs away with him and makes him a scientific Chauvin.

'Scientific Chauvinism, from which German men of science have hitherto kept themselves free, is more hateful than political, inasmuch as one expects decent demeanour more from scientific men than from politically excited masses. May it be far from us in the future also ! Let us not be misled in our intellectual habits by the present ebullition of national feeling in Europe. In spite of the tone of irritation appearing, now here, now there, among other nations, may we retain unlost the tradition of a scientific justice exercised without respect of nation, and of the serious literary work which this implies !

'May our Temple of the Muses remain a safe refuge for German cosmopolitanism if the storms of the time tolerate it nowhere else !'

Is not this conceived very much in the spirit of the well-known passage :—Ich danke dir, Gott, dass Ich nicht bin wie andere Leute, Räuber, Ungerechte, Ehebrecher ; oder auch wie dieser Zöllner ?

To any one who reads the above extract from Professor du Bois-Reymond's speech, it is obvious that the Chauvinism (surely Pharisaism would be the more correct word) so freely denounced (in others) towards the end, has been as freely practised (by the speaker himself) from the beginning.

But this special form of accusation is most particularly unhappy as directed against my book. For the book shows no Chauvinistic tendencies, properly so called:—its praise or blame may be deserved, or not, but they are certainly awarded from considerations altogether independent of nation or race; they are used throughout in favour of what I consider to be *true Science*, and against quackery, knavery, bigotry, and superstition, wherever found.

Fresnel and Carnot, Gauss and Riemann, Young and Faraday, are names to be honoured to all time; *not* by any means because they belonged to Frenchmen, Germans, or Britons; but because they belonged to men who have, each in his turn, led the van in the intellectual struggles of his generation.

But when a false prophet arises, or is raised up by others for the admiration of the unlearned multitude, it is a duty (often, it may be, a pleasant duty) to expose the hollowness of his pretensions; and to do so with sternly impartial relish whether he be French, German, or British. Equally is it a duty to bring forward the claims of a true prophet, be his nationality what it may; if these have suffered from his own modesty or carelessness, or from the neglect or disparagement of others.

My censor should have thought of the possible application of some of his own phrases to himself.

Was it not this fervent denouncer of Chauvinism who apologised to his students for the too Gallic sound of his own name? What but an absolutely overmastering antipathy to everything Gallic could have led a Professor of *Physiology* to speak of 'the fiery Celtic blood'—of a Norseman?

And the most recent authoritative text-book of *Spectrum Analysis*, published a year or two ago in Berlin, supplies a singular comment on the above eulogy of German scientific men in general. Though historical details are freely given in that work, the name of Balfour Stewart *is not even once mentioned!* I take this work as an example, because it is a high-class one. But, even from my own reading, which has been mainly confined to standard works (so far as German is concerned), I could supply numerous equally striking examples of exceptions to the sweeping statement so confidently made by my censor.

My acquaintance with Leibnitz's works may not be so profound as is that of Professor du Bois-Reymond ; but, such as it is, it has led me to accept the opinion of Huygens on him as a *man*, and that of Gauss on him as a *mathematician*. Surely even Professor du Bois-Reymond will allow that these (especially as neither was Gallic) were competent judges.

P. G. TAIT.

College, Edinburgh,
Dec. 29*th*, 1884.

PREFACE TO FIRST EDITION.

THE following Lectures were given in the spring of 1874, at the desire of a number of my friends,—mainly professional men,—who wished to obtain in this way a notion of the chief advances made in Natural Philosophy since their student days.

The only special requests made to me were, that I should treat fully the modern history of Energy, and that I should publish the Lectures *verbatim*.

The reader will judge for himself how far the first request has been attended to. As to the second, it is necessary to explain that, being very busy, I had not time to do more than arrange a few notes for each lecture ; so that the course was entirely extempore, and was taken down by excellent short-hand writers.

Besides necessary corrections, only one large change was made in the M.SS., viz., the excision of a great many of those repetitions which are indispensable in extempore lecturing, but are intolerable in a book. Professors Clerk-Maxwell and Balfour Stewart have been kind enough to read the proofs, and to suggest several valuable improvements.

The work must, however, be regarded as in no sense

whatever a finished production, though I hope it will be found not only accurate but also readable. In fact, I could not possibly have found time to rewrite the whole in the form in which I should like to have presented it for publication ; so that the reader is requested to remember, if he desires to find fault, that the non-removal of many defects whose correction would have required large changes, was the condition under which alone the book could have appeared. Still, I should not have allowed it to be published had I not been assured by competent judges that in spite of its necessary imperfections it is calculated to be useful.

P. G. TAIT.

COLLEGE, EDINBURGH,
February 1876.

CONTENTS.

LECTURE I.

INTRODUCTORY.

PAGE

Classification of Recent Advances in Physical Science. General Statement of the Objects of Physics. Time and Space. Matter, Position, Motion, and Force. Digression upon *à priori* reasoning. Instances of modern or revived fallacies—Uniformity of Earth's Rotation, Stability of Solar System, Heat developed the equivalent of work spent in compressing a gas, *Causa æquat effectum*. Gilbert the true originator of Experimental Science. Test of the reality of Matter—fails when applied to Force—not when applied to Energy. Conservation, Transformation, and Dissipation of Energy. Ignorance and Incapacity alike of Spiritualists and Materialists, ... 1

LECTURE II.

THE EARLY HISTORY OF ENERGY.

Newton's services to the subject only of late recognised. *Second Law* —There is no balancing of forces; but only of the effects of forces— *Geometrical* composition of velocities. *Third Law*—Its second interpretation an all but complete statement of the Conservation of Energy—*Arithmetical* composition of the squares of velocities. Experimental results of Rumford and Davy, filling up the *lacuna* in Newton's statement. Their proofs that Heat is not matter. Davy's statement of the true theory of Heat. Speculations of Séguin and Mayer, ... 27

LECTURE III.

ESTABLISHMENT OF THE CONSERVATION OF ENERGY.

Further inquiry into the asserted claims of Mayer. Opinions of Colding and Joule on Mayer's first paper. [Insertion (1884) on the prior claims of Mohr.] Colding's Experiments. Joule's Experiments. Numerical value of the Dynamical Equivalent of Heat. Helmholtz's argument from the Perpetual Motion. Transformation and Dissipation of Energy. Illustrative experiments, ... 52

LECTURE IV.

TRANSFORMATION OF ENERGY.

Experimental Illustrations—Heating of wires, and decomposition of water, by a Galvanic current—Electro-magnetic Engine—Rotating Disc—Magneto-electric Machine—Induction-Coil and Geissler Tube—Higher and Lower Forms of Energy. Work transformed wholly into Heat—Only a portion of the Heat can be reconverted into Work. Carnot's Cycle of Operations and his Reversible Cycle. Effect of pressure upon Ice, ... 81

LECTURE V.

TRANSFORMATION OF HEAT INTO WORK.

Carnot's Cycle—continued. Watt's Diagram of Energy. The Impossibility of *the* Perpetual Motion is an experimental truth. Conditions of Reversibility. Absolute definition of Temperature. Second Law of Thermodynamics. Absolute zero of temperature, or temperature of a body devoid of heat. Efficiency of the best steam-engine. Effect of pressure on the freezing point of water. Mechanism of Glacier motion, ... 107

LECTURE VI.

TRANSFORMATION OF ENERGY.

Further consequences of Carnot's ideas. Anomalous behaviour of water and of india-rubber. Application to rock masses, and the state of

CONTENTS.

the earth's interior. Availability of energy, and loss of availability. To restore the availability of one portion of energy, another portion must be degraded. Dissipation of energy. Sources of Terrestrial and of Solar Energy. Energy of plants and animals. Measure of the Sun's Radiant Energy. Energy now in the Solar System, . . 133

LECTURE VII.

SOURCES AND TRANSFERENCE OF ENERGY.

Available Sources of Energy on the Earth. Whence these have been derived. Uniformitarian School of Geologists. Sir W. Thomson's arguments as to the length of time during which life has been possible on the earth. Transference of Energy—through Solids, Fluids, and through the Ether. Test of the Receptivity of a body or system for energy in a vibratory form. Physical Analogies introductory to Spectrum Analysis, ... 162

LECTURE VIII.

RADIATION AND ABSORPTION.

History of the discovery of the Physical Basis of Spectrum Analysis. First result of Spectrum Analysis applied to non-terrestrial bodies;—There is Sodium gas in the Sun's Atmosphere. Elaborate experiments of Stewart and Kirchhoff. Identity of Light and Radiant Heat. Distinctive characters of a particular ray. Application of Carnot's principle to establish the equality of radiating and absorbing powers. Black, transparent, and perfectly reflecting bodies, . . . 187

LECTURE IX.

SPECTRUM ANALYSIS.

Spectrum of incandescent black body; of incandescent gas or vapour. Absorption by vapour of parts of spectrum of incandescent black body. Application to sunlight, and starlight. Solar spots and protuberances. Period of life of various stars. Fluorescence, . . 214

LECTURE X.

SPECTRUM ANALYSIS.

Change of colour of Light by relative velocity of source and observer. Analogy from sound. Causes of broadening of spectral lines. Spectrum of Solar Corona ; of Double Stars ; of Comets. Probable nature of Comets ; of Saturn's rings ; of the Zodiacal Light, . 237

LECTURE XI.

CONDUCTION OF HEAT.

Fourier's Mathematical Theory. His Definition of Conducting Power. Analogy between Thermal and Electric Conductivities. Forbes's method and results. Ångström's method. Penetration of Surface temperature into the earth's crust. Analogy between conduction of heat and conduction of electricity. Diffusion also analogous to these. Diffusion of matter, of kinetic energy, and of momentum, . . 265

LECTURE XII.

STRUCTURE OF MATTER.

Limits of Divisibility of Matter. In physics the terms great and small are merely *relative*. Various hypotheses as to structure of bodies —Hard Atom—Centres of Force—Continuous but Heterogeneous Structure—Vortex-atoms—[Digression on Vortex-Motion.] Lesage's Ultramundane Corpuscles. Proofs that matter has a grained structure. Approximation to its dimensions from the Dispersion of Light : —from the phenomena of Contact Electricity, . . 287

LECTURE XIII.

STRUCTURE OF MATTER.

Approximation to dimensions of grained structure from capillary phenomena—from properties of gases. Mathematical consequences of the supposition that a gas consists of constantly impinging particles—Gaseous Diffusion. Results of Maxwell's investigations. Physical reason of Dissipation—Andrews' results as to the continuity of the liquid and gaseous states of matter. Conclusion, . . 317

LECTURE XIV.

FORCE.

Evening Address to the British Association, Sept. 8, 1876, . . 343

LECTURE I.

INTRODUCTORY.

Classification of Recent Advances in Physical Science. General Statement of the Objects of Physics. Time and Space. Matter, Position, Motion, and Force. Digression upon *à priori* reasoning. Instances of modern or revived fallacies—Uniformity of Earth's Rotation, Stability of Solar System, Heat developed the equivalent of work spent in compressing a gas, *Causa æquat effectum.* Gilbert the true originator of Experimental Science. Test of the reality of Matter—fails when applied to Force—not when applied to Energy. Conservation, Transformation, and Dissipation of Energy. Ignorance and Incapacity alike of Spiritualists and Materialists.

In considering what may be designated as 'Recent Advances in Physical Science,' it is well to remember that many things which have become almost popularly known within the last twenty-five years are much older in the minds and writings of the foremost scientific men. We cannot, however, treat them intelligibly without reference, sometimes pretty full, to what was known even earlier still: so that you must not be surprised if I have a good deal to say of Davy and Rumford, and even of Newton.

I shall, for the sake of clearness, attempt roughly to classify these recent advances under five well-marked heads; but I shall do so very briefly, deferring explanation even of new scientific terms till I have to treat each of these heads in detail.

First and foremost, advances connected with the

modern notion of Energy. Just as Gold, Lead, Oxygen, etc., are different kinds of Matter, so Sound, Light, Heat, etc., are now ranked as different forms of Energy, which, as we shall presently see, has been shown to have as much claim to objective reality as matter has. This grand idea enables us to co-ordinate all the parts, however apparently diverse, of the enormous subject of Natural Philosophy. It has not only thus enabled us to exhibit the science in a complete and connected form, but it has also, specially by the application of the laws of Thermo-dynamics (to which a large part of this course will be devoted), enabled us to find those points where rapid advance was most easily to be secured.

Secondly. The advances which have arisen, more or less directly, from the requirements felt in practical applications. To take but a single instance: think of the immense improvements in instruments for the measurement of electric charges and electric currents, such as electrometers and galvanometers, which have been effected because called for by the recent extensions of submarine telegraphy. It is not too much to say that the instruments now employed, and which were primarily devised for practical telegraphic purposes, are hundreds of times more sensitive, as well as more exact, and therefore more useful for purely scientific purposes, than the best of those which were in use thirty years ago. Thus it is that a development of science, in a practical direction, leads to the construction of instruments which have, as it were, a reflex action on the development of the pure science itself.

Thirdly. Those which arise from the assistance rendered to one another by pure sciences, such as astronomy, chemistry, and physiology, where, in fact, the improvement of one branch has led, almost immediately, to important extensions of other branches. Under this head we may also include those very great advances which are due to improvements in our mathematical methods.

Fourthly. What may be called casual discoveries, though they are often of very great importance; such as, for instance, the discovery of fluorescence, with its manifold consequences, and the invention of the processes of photography. Such discoveries, instead of being, as in old days, wondered at and left isolated, are now at once attacked on all sides by numberless enthusiastic experimenters.

Fifthly. There is another class, very numerous but more difficult to exactly describe. As a single example of this class, I may mention the modern statistical methods of treating certain problems of physical science, especially those connected with the movements of particles of gases and liquids, to which I shall advert at considerable length in the course of these lectures.

I have now to consider how I should best commence the analysis of these various heads; and I think the proper method will be first to sketch the subject as if from a distance—to point out a few of the principal peaks which we have to ascend, and of the more formidable abysses which we have to avoid; striving all the while to introduce as early as possible some of those new technical terms which are absolutely indispensable to accuracy and definiteness, and which, therefore, can not be too soon mastered.

Natural Philosophy, as now regarded, treats generally of the physical universe, and deals fearlessly alike with quantities too great to be distinctly conceived, and with quantities almost infinitely too small to be perceived even with the most powerful microscopes; such as, for instance, distances through which the light of stars or nebulæ, though moving at the rate of about 186,000 miles per second, takes many years to travel; or the size of the particles of water, whose number in a single drop may, as we have reason to believe, amount to somewhere about

10^{26}, or 100,000,000,000,000,000,000,000,000.

Yet we successfully inquire not only into the composition of the atmospheres of these distant stars, but into the number and properties of these water-particles, nay, even into the laws by which they act upon one another.

The fundamental notions which occur to us when we commence the study of physical science are those of *Time* and *Space*. A measure of time may be obtained by physical methods, as in fact is done incidentally in Newton's First Law of Motion, wherein he asserts that a mass left to itself moves *uniformly*. That is, equal times are the times in which such a mass describes equal spaces. Of space, we can ascertain by observation the properties. But we cannot inquire into the actual nature of either space or time, except in the way of a purely metaphysical, and therefore of necessity absolutely barren, speculation. We have, however, mathematical methods specially adapted to the treatment of these two abstract ideas; Algebra, which has been called (by Sir W. R. Hamilton) the science of pure time; and Geometry, which may be designated the science of pure space.

The common measurement of time primarily depends upon the rotation of the earth about its axis. This, however, as will be seen when we advance a little further, is by no means a uniform quantity, and therefore ultimately the measurement of time must be based upon some motion depending on a physical property of matter which we have every experimental reason for believing to be unchangeable by time, and invariable throughout the universe. Probably such an ultimate standard for the measurement of time will be found in one of the periods of vibration of the molecules of a heated gas, such as hydrogen, under given conditions.

The properties of space, involving (we know not why) the essential element of three dimensions, have recently been subjected to a careful scrutiny by mathematicians of the highest order, such as Riemann and Helmholtz;[1] and the result of their inquiries leaves it as yet undecided whether space may or may not have precisely the same properties throughout the universe. To obtain an idea of what is meant by such a statement, consider that in crumpling a leaf of paper, which may be taken as representing space of two dimensions, we may have some portions of it plane, and other portions more or less cylindrically or conically curved. But an inhabitant of such a sheet, though living in space of two dimensions only, and therefore, we might say beforehand, incapable of appretiating the third dimension, would certainly feel some difference of sensations in passing from portions of his space which were less, to other portions which were more, curved. So it is possible that in the rapid march of the solar system through space, we may be gradually passing to regions in which space has not precisely the same properties as we find here—where it may have something in three dimensions analogous to curvature in two

[1] See Helmholtz' paper in *Mind*, No. III. 1876.

dimensions—something, in fact, which will necessarily imply a fourth-dimension change of form in portions of matter in order that they may adapt themselves to their new locality. But for the full discussion of a question like this it would be necessary to introduce mathematical reasoning of a transcendental character.

In addition to these fundamental notions of time and space, the next four which force themselves upon us in the physical universe are those of Matter, Position, Motion, and Force. As with these ideas commences the study of physics proper, I leave them for a moment to consider in what way or in what spirit we ought to treat problems of physical science. Remember that the subject of my lectures is the *Advances* of Physical Science. It is well then to inquire briefly to what we are indebted for such advances. And every one who has with any attention studied the history of scientific progress sees at once that

These advances come or not according as we remember or forget that our science is to be based entirely upon experiment or mathematical deductions from experiment.

There is nothing physical to be learned *à priori*. We have no right whatever to ascertain a single physical truth without seeking for it physically, unless it be a necessary consequence of other truths already acquired by experiment, in which case mathematical reasoning is alone requisite.

Let us consider for a moment to what fearfully absurd consequences a neglect of this self-evident principle has led in former times, and too often even in modern days. Men were told by the antients that the planets move in circles because circular motion is perfect! They were told also in the middle ages that the sun cannot possibly have spots! They were told that the earth was at rest; that Nature abhors a vacuum, etc. etc. And all these dogmas were enuntiated by otherwise reasonable men. Within the last fifty years we have had philosophers like Hegel saying that the motion of the heavenly bodies is not a being pulled this way and that: that they go along, as the antients said, like blessed gods. Further, that pressure, gravity, etc., are true only of terrestrial, not of celestial matter. Hegel winds up this truly wonderful statement by saying that both are matter, just as a good thought and a bad one are both thoughts, but the bad is not therefore good because the good one is a thought.[1]

As instances of still more recent, in fact quite modern, fallacies of a somewhat similar kind, I shall take but four, two of which are in their very nature excusable,— the other two utterly unpardonable.

First, there is the assumption that the earth's rotation is absolutely uniform. Now, to say nothing of the effects of cooling and consequent shrinking, the effects of volcanic disturbances and upheavals, the effects of degradation of mountains, and various other causes

[1] *Naturphilosophie*, § 269. [The passage is so incredibly absurd that I feel bound to quote it.] Die Bewegung der Himmelskörper ist nicht ein solches Hin- und Hergezogenseyn, sondern die freie Bewegung; sie gehen, wie die Alten sagten, als selige Götter einher. Die himmlische Körperlichkeit ist nicht eine solche, welche das Princip der Ruhe oder Bewegung ausser ihr hätte. Weil der Stein träge ist, die ganze Erde aber aus Steinen besteht, und die andern himmlischen Körper eben dergleichen sind—ist ein Schluss, der die Eigenschaften des Ganzen denen des Theils gleichsetzt. Stoss, Druck, Widerstand, Reibung, Ziehen und dergleichen gelten nur von einer andern Existenz der Materie, als die himmlische Körperlichkeit. Das Gemeinschaftliche Beider ist freilich die Materie, so wie ein guter Gedanke und ein schlechter beide Gedanken sind: aber der schlechte nicht darum gut, weil der gute ein Gedanke ist.

which must tend more or less to affect the earth's rotation (shrinking and degradation accelerating it, while upheavals retard it, according to a mechanical principle which is involved in Newton's *Third* Law of Motion), there has been recently revived the study, first pointed out by Kant, of the effect of tidal retardation upon the length of the day. In fact, the earth with the tide-wave upon it, pointing on the average almost axially towards the moon, is virtually revolving in a friction-brake or collar; and so long as it moves with reference to this tidal wave, so long must it move subject to friction, and therefore of course with continually decreasing velocity.

Then, again, we had the confident assertion of the absolute stability of the solar system; that is to say, grand arguments were founded by the Teleologists on the assumption that the eccentricities and inclinations, and so on, of the planetary orbits, though constantly varying, fluctuated between certain definite, and in general very narrow limits, and that after a by no means long series of ages all bodies in the solar system would return to almost precisely their former configuration as to position and velocity. Now, in arriving at this result, which of course they themselves understood in its true sense, Laplace and Lagrange confessedly employed approximate methods of solution only. They left out of account what are termed technically the squares of disturbing forces; that is to say, of two planets, each of which has disturbed the other's position, the effects of the first upon the second were calculated by leaving out of account the disturbance of the position of the first, and *vice versâ*. In order to improve upon this approximation, at least without enormous labour, mathematical methods of a far more powerful order than have yet been invented are requisite, and therefore it is not from this point of view that the solution can at present be improved ; nor can we well form an idea of the nature of the modification which the results of the approximate method would undergo. But the idea which I have just mentioned with reference to tidal friction, which has not yet been taken account of in the solution of these planetary problems, shows at once that so long as the parts of any moving integral portion of the system are capable of being displaced relatively to one another, and so moving relatively with friction, so long must there be a cause tending constantly to the degradation of the rates of motion in the system, and therefore that stability of the planetary system is impossible under present conditions. Remember that it was in the imagined interests of religion that the earth's motion was denied. History repeats itself here. An ill-informed Teleologist, however good his intentions, is far more dangerous to the cause he has at heart than the bitterest of its declared enemies.

Then let us take the question of the heat developed by compressing a gas. You all know that a piece of tinder can be set on fire when it is enclosed in a cylinder in which the air is suddenly compressed by pushing in a tight-fitting piston. Great credit has recently been claimed for two speculators, Séguin and Mayer, who independently propounded the hypothesis that the heat developed in such a case is the equivalent of the work spent in compressing the air ; or its converse, that the heat lost in expansion is the equivalent of the work done by the expanding material. To make such hypotheses without preliminary experimental measurements, is simply to fall into the fatal error to which I have already adverted,—the *à priori* assertion of physical principles. To see that it is so, we have only to consider that a gas might (for all we can tell without experiment) have the properties of a spiral spring. Suppose, in fact, instead of air, the cylinder above spoken of to be filled with a number of spiral springs so adjusted as not to interfere with each other's motions. In compressing such a set of springs, exactly the same amount of work may be spent as in compressing air, and yet we may find no trace whatever of heat generated. It therefore appears obvious that until we know for certain the ultimate nature of a gas, the only way (independent of mere guessing) to discover the relation between the heat developed by compression and the work spent in producing it, is to experiment; and that without experiment it is impossible to lay down any general relation between them. The modern view of the constitution of a gas, in which its particles are supposed to be flying about with great velocity in all directions, and constantly impinging upon one another and upon the sides of the vessel, leads us almost directly to many valuable conclusions, among which I will refer for the moment only to the result known as Boyle's law, where we contemplate the compression of a gas whose temperature is kept constant. Suppose, for instance, the particles to be moving with a certain velocity in every direction, we find that if the piston could be moved half way down the cylinder, and the velocity of the particles not thereby increased,[1] the number of impacts per second upon the ends of the cylinder must become twice as great as it was before,

[1] This would be a violation of the principle of Dissipation of Energy, as will be seen by the reader of Lecture VI. But that does not invalidate its usefulness as an illustration of the present argument.

because the length of the cylinder is only half as great. Also, the number of impacts per second per square inch upon the curved sides of the cylinder must likewise be doubled, simply because there is the same number of particles as before, impinging with the same velocities, but upon only one half of the surface. If we could manage to advance the whole piston by infinitesimally small stages, so as at each such advance to take advantage of the absence of all molecular pressure upon the piston, or to advance at every instant those parts of the piston upon which for the moment no impact was impending, we should produce this diminution of bulk without altering in any respect the velocities of the particles of gas; and therefore, according to Boyle's law, and according to the analysis just given, we should have the case of a gas doubled in pressure, and occupying exactly one half the bulk which it occupied at first, but without increase of temperature. Here then is another mode of contemplating the compression of a gas without any production of heat. This question is one of great importance, and I intend to treat it pretty fully in the course of these lectures.

The only other fallacy which I shall mention for the present, is that of basing physical results upon the old dog-Latin dogma, *causa æquat effectum*. It is difficult to decide whether the Latinity or the (semi-obscure) sense is in this dogma the more incorrect. The fact is, that we have not yet quite cast off that tendency to so-called metaphysics which has often completely blasted the already promising career of a physical inquirer. I say 'so-called' metaphysics, because there is a science of metaphysics; but from the very nature of the case, the professed metaphysicians will never attain to it. In fact if we once begin to argue upon such a dogma as the above, the next step may very naturally be to inquire whether cause and effect are simultaneous or successive :—and then we shall have become so mystified about the meaning of the word Cause that we may well be ready to inquire (as many have already done) what is the necessarily ever acting cause of the uniform motion of a body upon which no forces act!

The originator of true experimental science seems to have been Gilbert of Colchester, whose deservedly celebrated treatise *De Magnete* was published 300 years ago. After him came Galileo and Newton, each making gigantic strides in the true direction, and by them this, the ONLY way of attaining to a discovery of physical laws, was permanently established. The proof of this is, that the last two centuries and a half have achieved, in purely physical science, million-fold what had been accomplished before them. And it is not that we are now more able, nor that we have more leisure—certainly not :—

'. . . for Romans now
Have thewes and limbs like to their ancestors'.'

It is rather that whenever the direction given to inquiry is a proper one, the men come forward. This direction was good in Britain at certain memorable times, as when Newton and Hooke were contemporaries ; in the days of Maclaurin and Cotes, and in those of Cavendish and Watt. At intervals it broke down entirely as regards mathematical physics, partly as regards experimental physics, and once again it has become good ; and consequently, since the ever-memorable days of Young and Davy, we have had Green and Hamilton, Faraday and Graham, and we can still rejoice in the possession of

Stokes and Thomson, Adams and Clerk-Maxwell, Joule and Andrews. This list is as good as either of the others, and might be considerably increased. Other countries have had their similar fluctuations, all I believe traceable to similar causes. Little more than half a century ago, France had such mighty names as Ampère, Laplace, Lagrange, Poisson, Fresnel, Fourier, Carnot, Cauchy, etc. I name them just as they occur to me. We cannot do much in the way of classifying men like these. Germany now has Helmholtz, Weber, Kirchhoff, and has but recently lost Gauss, Jacobi, Dirichlet, Plücker, Riemann, and Magnus.

The sad fate of Newton's successors ought ever to be a warning to us. Trusting to what he had done, they allowed mathematical science almost to die out in this country, at least as compared with its immense progress in Germany and France. It required the united exertions of the late Sir J. Herschel and many others to render possible in these islands a Boole and a Hamilton. If the successors of Davy and Faraday pause to ponder even on *their* achievements, we shall soon be again in the same state of ignominious inferiority. Who will then step in to save us?

Even as it is, though we have among us many names quite as justly great as any that our rivals can produce, we have also (even in our educated classes) such an immense amount of ignorance and consequent credulity, that it seems matter for surprise that true science is able to exist. Spiritualists, Circle-squarers, Perpetual-motionists, Believers that the earth is flat and that the moon has no rotation, swarm about us. They certainly multiply much faster than do genuine men of science. This is characteristic of all inferior races, but it is consolatory to remember that in spite of it these soon become extinct. Your quack has his little day, and disappears except to the antiquary. But in science nothing of value can ever be lost; it is certain to become a stepping-stone on the way to further truth. Still, when our stepping-stones are laid, we should not wait till others employ them. 'Gentlemen of the Guard, be kind enough to fire first,' is a courtesy entirely out of date; with the weapons of the present day it would be simply suicide.

To come back to our second set of elementary ideas, Matter, Position, Motion and Force. Of these, the second (Position) is a purely space relation, or geometrical conception, and must necessarily be relative, unless something like the idea of Riemann already referred to have an actual existence in the universe. The third (Motion) is mere change of position, but as that change may take place more or less rapidly, it involves the idea of time as well as of space. But both of these ideas are quite independent of the remaining two (Matter and Force); and in fact their study forms the subject of a special mixed science of Time and Space, called *Kinematics*, which takes its place beside the older sciences, Geometry and Algebra, which I have already adverted to as the sciences of pure Space and pure Time.

The grand test of the reality of what we call Matter, the proof that it has an objective existence, is its indestructibility and uncreateability—if the term may be used—by any process at the command of man. The value of this test to modern chemistry can scarcely be estimated. In fact we can barely believe that there could have existed an exact science of chemistry had it not been for the early recognition of this property of matter; nor in fact would there be the possibility of a chemical analysis, supposing that we had not the assurance by enormously extended series of previous experiments, that no portion of matter, however small, goes out of existence or comes into existence in any operation whatever. If the chemist were not certain that at the end of his operations, provided he has taken care to admit nothing and to let nothing escape, the contents of his vessels must be precisely the same in quantity as at the beginning of the experiment, there could be no such thing as chemical analysis. Some substance might suddenly appear,[1] or some substance might suddenly vanish, and no reasoning whatever could lead to a deduction from the results of experiments under such conditions. This, then, is to be looked upon as the great test of the objective reality of matter.

There remains to be treated Force, the last of the fundamental four. The notion is suggested to us directly, by the so-called 'muscular sense,' which gives us the feeling of pressure, as when we move a body with our hand or foot. But we must be particularly cautious as to the way in which we treat the evidence of our senses in such matters. Think of Sound and Light, for instance—which, till they affect a special organ of sense, are mere wave-motions. The sensation is as different from the cause in such cases as are the bruise and the

---

[1] Hegel believed in such possibilities. Witness, among others, the following—almost the raciest of the manifold absurdities of the *Naturphilosophie*. It occurs in § 332. Ebenso werden die kaustischen Kali wieder milde; man sagt dann, sie ziehen Kohlensäure aus der Luft ein. Das ist aber eine Hypothese; sie machen vielmehr aus der Luft erst Kohlensäure, um sich abzustumpfen.

pain produced by a cudgel or a cricket ball from the mere motion of those portions of matter before impact on a part of the human body. In all likelihood a similar (probably a more sweeping) statement is true of force. [This subject is treated in a special Lecture, appended to the present work.]

The definition of force in physical science is implicitly contained in Newton's First Law of Motion, and may thus be given:—

Force is any cause which alters a body's natural state of rest or of uniform motion in a straight line.

The only difficulty, and it is a serious one, which we feel here, is as to the word 'cause;' for this, amongst material things, usually implies objective existence. Now we have absolutely no proof of the objective existence of force in the sense just explained. In every case in which force is said to act, what is really observed, independent of the muscular sense (whose indications, like those of the sense of touch in matters concerning the temperatures of bodies, are apt to be excessively misleading), is either a transference, or a tendency to transference, of what is called energy from one portion of matter to another. Whenever such a transference takes place, there is relative motion of the portions of matter concerned, and the so-called force in any direction is merely the rate of transference, or of transformation, of energy per unit of length for displacement in that direction. Force then has not necessarily objective reality any more than has Velocity or Position. The idea, however, is still a very useful one, as it introduces a term which enables us to abbreviate statements which would otherwise be long and tedious; but, as Science advances, it is in all probability destined

INTRODUCTORY.

to be relegated to that Limbo which has already received the Crystal Spheres of the Planets, and the Four Elements, along with Caloric and Phlogiston, the Electric Fluid and the Odic or Psychic Force.

It is only, however, within comparatively recent years that it has been generally recognised that there is something else in the physical universe which possesses to the full as high a claim to objective reality as matter possesses, though it is by no means so tangible, and therefore the conception of it was much longer in forcing itself upon the human mind. The so-called 'imponderables,'—things of old supposed to be matter—such as heat and light, *et cetera*, are now known by the purely experimental, and therefore the only safe, method to be but varieties of what we call Energy,—something which, though not matter, has as much claim to recognition on account of its objective existence as any portion of matter. The grand principle of Conservation of Energy,[1] which asserts that no portion of energy can be put out of existence, and no amount of energy can be brought into existence by any process at our command, is simply a statement of the invariability of the quantity of

---

[1] Great confusion has been introduced into many modern British works by a double use of the word Force. It is employed, without qualification, sometimes in the sense of force proper (as above defined), sometimes in the sense of energy! The two things (if force proper can be called a 'thing,' having probably no objective existence, and certainly no conservation, except possibly in a highly refined sense, which Faraday in vain attempted to realise experimentally, but which, even if it were proved, would have no connection with conservation of energy) are of as different orders as miles and square miles, though perhaps they are not quite so incomparable as minutes and yards or pence. Even a mere want of precision in the use of terms of such fundamental importance is altogether incompatible with the existence of true scientific method. [See Lecture XIV. (on Force) at the end of this volume.]

energy in the universe,—a companion statement to that of the invariability of the quantity of matter.

The laws of energy differ from those of matter in one most important respect, so far at least as we yet know by experiment. Matter cannot, so far as we yet know, be transmuted from one kind to another, though in some cases it assumes what is called an *allotropic* form. The great characteristic of energy, on the other hand, is that in general we can readily transform it (in fact it is of use to us solely because it can be transformed), but in all its transformations the quantity present remains precisely the same.

Energy may be defined as the power of doing work, or, if we like to put it so, of doing mischief. I have already pointed out to you that the notion of energy is harder to seize than that of matter. Wherein, for instance, consists the difference between a mass of snow lying on the mountain side and the same mass when it has fallen and rests in the valley below? Obviously, so far as the matter present is concerned, the two subtances are identical, except in so far as molecular changes, such as melting, may have altered the state of some portions of the mass during or after its descent. Yet the elevated mass possesses, in virtue of its elevation alone, a power of doing work or mischief, which it has lost entirely when it has descended as far as it can. By the mere fact, then, of its elevation, it possesses a power which it does not possess when it has descended. This is called energy of position, or *Potential Energy*. Other examples of it are to be found in a wound-up spring or weight, as in a clock, a bent bow; or in gunpowder; and various others might easily be mentioned. Perhaps the most striking of all instances that we can give is that of the food of animals, including as one of the principal constituents the oxygen of the atmosphere.

But when the snow is detached from the mountain side, in descending it acquires another form of energy, depending entirely on its motion; and thus we distinguish between energy of position and energy of motion or *Kinetic Energy*. To those who have acquired the intelligent use of the terms it is matter of common observation that as the one of these quantities becomes less, the other becomes greater. The velocity of the falling snow increases constantly as it gradually descends; and exact calculation, according to physical experiment, shows us that the amount of potential energy lost in every stage of the operation is precisely equal to the amount of Kinetic energy gained. The process may be inverted if we consider Kinetic energy to be originally communicated to a body, suppose, for simplicity, in a vertically upward direction. We know that a stone thrown into the air gradually loses velocity as it ascends higher and higher; for an instant, when it has lost all velocity, it pauses, and then returns, gradually regaining velocity, as it in turn loses its advantage of position; and calculation, applied to this case, shows that at every stage, whether of the ascent or of the descent, the sum of the Potential and the Kinetic energies remains precisely the same, except in so far as it is modified by the resistance of the air. This, however, gives us no exception to the general truth of the principle of conservation of energy, because any energy lost by the stone is communicated without loss of quantity to the surrounding air.

We contemplate, therefore, with reference to energy, its conservation, which merely asserts its objective reality; its transformations, which render it indispensable to the existence of life and the physical changes in the universe; but it has in addition another and even more curious property. We have seen that change is essential to the existence of phenomena such as we observe: and, that this change may take place, it is necessary that there should be constant transformations of energy. But some forms of energy are more capable of being transformed than others; and every time that a transformation takes place, there is always a tendency to pass, at least in part, from a higher or more easily transformable to a lower or less easily transformable form.

Thus the energy of the universe is, on the whole, constantly passing from higher to lower forms, and therefore the possibility of transformation is becoming smaller and smaller, so that after the lapse of sufficient time all higher forms of energy must have passed from the physical universe, and we can imagine nothing as remaining, except those lower forms which are incapable, so far as we yet know, of any further transformation. The low form to which all transformations with which we are at present acquainted seem inevitably to tend, is that of uniformly diffused heat: or, more precisely, heat so diffused as to produce uniform temperature. We know, in fact, that in order to make any use of heat —to transform it into mechanical power or into any other form of energy—it is absolutely necessary that we should have bodies of different temperatures. We must, as it were, have a source and a condenser. Now, when all the energy of the universe has taken the final form of heat so diffused as to produce uniform temperature, it will obviously be impossible to make any use of this heat for further transformation. Thus, so far as we can as yet determine, in the far distant future of the universe the quantities of matter and energy will remain absolutely as they now are—the matter unchanged alike in quantity and quality, but collected together under the influence of its mutual gravitation, so that there remains no potential energy of detached portions of matter; the energy also unchanged in quantity, but entirely transformed in quality to the low form of heat so diffused as to produce uniformity of temperature.[1]

This, the Dissipation of Energy,[2] is by no means well understood, and many of the results of its legitimate application have been received with doubt, sometimes even with attempted ridicule. Yet it appears to be at the present moment by far the most promising and fertile portion of Natural Philosophy, having obvious applications of which as yet only a small percentage appear to have been made. Some, indeed, were made before the enuntiation of the Principle, and have since been recognised as instances of it. Of such we have good examples in Fourier's great work on Heat-conduction, in the optical theorem that an image can never be brighter than the object, in Gauss's mode of investigating electrical distribution, and in some of Thomson's theorems as to the energy of an electromagnetic field. But its discoverer has, so far as I know, as yet confined himself in its explicit application to questions of Heat-conduction and Restoration of Energy, Geological Time, the Earth's Rotation, and such like. Unfortunately his long-expected *Rede Lecture*[3] has not yet been published, and

[1] Thomson *On a Universal Tendency in Nature to Dissipation of Energy.* Proc. R.S.E. 1852.
[2] What follows is extracted from my address as President of Section A at the British Association Meeting of 1871.
[3] Delivered in the Senate House, Cambridge, in 1866.

its contents (save to those who were fortunate enough to hear it) are still almost entirely unknown.

But there can be little question that the Principle contains implicitly the whole theory of Thermo-electricity, of Chemical Combination, of Allotropy, of Fluorescence, etc., and perhaps even of matters of a higher order than common physics and chemistry. In Astronomy it leads us to the grand question of the *age*, or perhaps more correctly the *phase of life*, of a star or nebula, shows us the material of potential suns, other suns in the process of formation, in vigorous youth, and in every stage of slowly protracted decay. It leads us to look on each planet and satellite as having been at one time a tiny sun, a member of some binary or multiple group, and even now (when almost deprived, at least at its surface, of its original energy) presenting an endless variety of subjects for the application of its methods. It leads us forward in thought to the far-distant time when the materials of the present stellar systems shall have lost all but their mutual potential energy, but shall in virtue of it form the materials of future larger suns with their attendant planets. Finally, as it alone is able to lead us, by sure steps of deductive reasoning, to the necessary future of the universe—necessary, that is, if physical laws for ever remain unchanged—so it enables us distinctly to say that the present order of things has *not* been evolved through infinite past time by the agency of laws now at work, but must have had a distinctive beginning, a state beyond which we are totally unable to penetrate; a state, in fact, which must have been produced by other than the now [visibly] acting causes.

Thus also it is possible that in Physiology it may, ere

long, lead to results of a different and much higher order of novelty and interest than those yet obtained, immensely valuable though these certainly are.

It was a grand step in science which showed that just as the consumption of fuel is necessary to the working of a steam-engine, or to the steady light of a candle, so the living engine requires food to supply its expenditure in the forms of muscular work and animal heat. Still grander was Rumford's early anticipation that the animal is a more economic engine than any lifeless one we can construct. Even in the explanation of this there is involved a question of very great interest, still unsolved, though Joule and many other philosophers of the highest order have worked at it. Joule has given a suggestion of great value, viz., that the animal resembles an electromagnetic- rather than a heat-engine ; but this throws us back again upon our difficulties as to the nature of electricity. Still, even supposing this question fully answered, there remains another—perhaps the highest which the human intellect is capable of directly attacking, for it is simply preposterous to suppose that we shall ever be able to understand scientifically the source of Consciousness and Volition, not to speak of loftier things—there remains the question of Life. Now it may be startling to some of you, especially if you have not particularly considered the matter, to hear it surmised that possibly we may, by the help of physical principles, especially that of the Dissipation of Energy, some time attain to a notion of what constitutes Life—mere Vitality, I repeat, nothing higher. If you think for a moment of the vitality of a plant or a zoophyte, the remark perhaps will not appear so strange after all. But do not fancy that the Dissipation of Energy to which I refer is at all that of a watch or suchlike piece of mere human mechanism, dissipating the low and common form of energy of a single coiled spring. It must be such that every little part of the living organism has its own store of energy constantly being dissipated, and as constantly replenished from external sources drawn upon by the whole arrangement in their harmonious working together. As an illustration of my meaning, though an extremely inadequate one, suppose Vaucanson's Duck to have been made up of excessively small parts, each microscopically constructed, as perfectly as was the comparatively coarse whole, we should have had something barely distinguishable, save by want of instincts, from the living model. But let no one imagine that, should we ever penetrate this mystery, we shall thereby be enabled to produce, except from life, even the lowest form of life. Sir W. Thomson's splendid suggestion of Vortex-atoms, if it be correct, will enable us thoroughly to understand matter, and mathematically to investigate all its properties. Yet its very basis implies the *absolute necessity* of an intervention of Creative Power to form or to destroy one atom even of dead matter. The question really stands thus :—Is Life physical or no? For if it be in any sense, however slight or restricted, physical, it is to that extent a subject for the Natural Philosopher, and for him alone.

There must always be wide limits of uncertainty (unless we choose to look upon Physics as a necessarily finite Science) concerning the exact boundary between the Attainable and the Unattainable. One herd of ignorant people, with the sole *prestige* of rapidly increasing numbers, and with the adhesion of a few fanatical deserters from the ranks of Science, refuse to admit that all the phenomena even of ordinary dead matter are strictly and exclusively in the domain of physical science. On the other hand, there is a numerous group, not in the slightest degree entitled to rank as Physicists (though in general they assume the proud title of Philosophers), who assert that not merely Life, but even Volition and Consciousness are merely physical manifestations. These opposite errors, into neither of which it is possible for a genuine scientific man to fall, so long at least as he retains his reason, are easily seen to be very closely allied. They are both to be attributed to that Credulity which is characteristic alike of Ignorance and of Incapacity. Unfortunately there is no cure; the case is hopeless, for great ignorance almost necessarily presumes incapacity, whether it show itself in the comparatively harmless folly of the Spiritualist or in the pernicious nonsense of the Materialist.

Alike condemned and contemned, we leave them to their proper fate—oblivion ; but still we have to face the question, where to draw the line between that which is physical and that which is utterly beyond physics. And, again, our answer is—Experience alone can tell us ; for experience is our only possible guide. If we attend earnestly and honestly to its teachings, we shall never go far astray. Man has been left to the resources of his intellect for the discovery not merely of physical laws, but of how far he is capable of comprehending them. And our answer to those who denounce our legitimate studies as heretical is simply this,—A revelation of anything which we can discover for ourselves, by studying the ordinary course of nature, would be an absurdity.

A profound lesson may be learned from one of the earliest little papers of Sir W. Thomson, published while he was an undergraduate at Cambridge, where he shows that Fourier's magnificent treatment of the Conduction of Heat [in a solid body] leads to formulæ for its distribution which are intelligible (and of course capable of being fully verified by experiment) for all time future, but which, except in particular cases, when extended to time past, remain intelligible for a finite period only, and *then* indicate a state of things which could not have resulted under known laws from any conceivable previous distribution [of heat in the body]. So far as heat is concerned, modern investigations have shown that a previous distribution of the *matter* involved may, by its potential energy, be capable of producing such a state of things at the moment of its aggregation; but the example is now adduced not for its bearing on heat alone, but as a simple illustration of the fact that all portions of our Science, and especially that beautiful one, the Dissipation of Energy, point unanimously to a beginning, to a state of things incapable of being derived by present laws [of tangible matter and its energy] from any conceivable previous arrangement.

I conclude by quoting some noble words used by Stokes in his Address to the British Association at Exeter :—' When from the phenomena of life we pass on to those of mind, we enter a region still more profoundly mysterious. . . . Science can be expected to do but little to aid us here, since the instrument of research is itself the object of investigation. It can but enlighten us as to the depth of our ignorance, and lead us to look to a higher aid for that which most nearly concerns our wellbeing.'

LECTURE II.

THE EARLY HISTORY OF ENERGY.

Newton's services to the subject only of late recognised. *Second Law*—There is no balancing of forces; but only of the effects of forces—*Geometrical* composition of velocities. *Third Law*—Its second interpretation an all but complete statement of the Conservation of Energy—*Arithmetical* composition of the squares of velocities. Experimental results of Rumford and Davy, filling up the *lacuna* in Newton's statement. Their proofs that Heat is not matter. Davy's statement of the true theory of Heat. Speculations of Séguin and Mayer.

THOUGH the subject which has been proposed to me is, 'The Advances of Physical Science within the last thirty years,' we must look upon the calling attention to valuable though neglected or misunderstood discoveries of old time, as being quite as much an advance in the present age as anything that has been done for the first time within the last few years. I cannot commence better than with those two of the great advances made by Newton, which were unfortunately very little recognised during his life, but which within the last ten or twelve years have been brought prominently before the world, and have shown us how enormously in advance of his time—and perhaps in some respects even of our time—Newton was.

The first of these is contained in his simple statement of the Second Law of Motion. I shall read it, not in his own words, but in a translation. He says:

'*Change of motion is proportional to force, and takes place in the direction of the straight line in which the force acts.*' Now, for the century and a half since Newton's time, mathematicians and natural philosophers have been puzzling themselves to invent various proofs—so-called statical proofs—of the law of composition of forces; the law which informs us how we are to find a single force which will produce precisely the same effect upon a body as two simultaneously acting forces applied at one point. All these different schemes have been, I may say, one more complex than another; and they have finally landed the student in utter confusion. Out of that confusion we have only recently escaped by coming back to the simple, but extraordinarily complete, statement of Newton's which I have just read.

Newton tells you, 'Change of motion is proportional to force.' He says nothing whatever as to what the motion was to begin with. He says nothing whatever as to the force being alone. There may be as many forces acting as we please; and of every one of them he says the change of motion which it produces is proportional to it, and takes place in its direction.

Moreover, in that statement Newton tells us that a force, according to him, always produces an effect. There is no such thing as two or more forces balancing one another—preventing one another from acting, as it were. Newton's notion is, if there is a force at all, it is doing something; and what it does is, it produces a change of motion, or, in modern language, a change of momentum, proportional to itself and in its own direction. So that, according to Newton, there is practically no such thing as Statics. There is no balancing of forces. There is balancing of the effects of forces, which is quite another thing. A force always produces its effect, and if two forces or more produce effects which balance one another, then we shall have perpetual balancing; but we have no balancing forces, merely a balancing of the effects they produce. We have the very simplest case of this where a weight is lying on a table. Gravity is constantly acting : the weight is constantly being pulled down by the attraction of the earth, but it is as constantly being pressed upwards by the resistance of the table; and each of these is producing in each second a certain quantity of momentum. The one is producing momentum in a vertically downward direction; the other is producing momentum in a vertically upward direction. These correspond to equal velocities in an upward and a downward direction; but it is the velocities, not the forces, which balance or neutralise one another.

To extend this statement to the case of the fundamental proposition in statics, which tells us how to compound two forces, and to find their resultant, all we have to do is to consider the two forces as acting upon a single particle of matter. If one of them acted alone, for a certain time, it would give it a velocity of a certain amount, and in a certain direction. If the other acted alone, for the same period of time, it would equally give a velocity definite in amount, and definite in direction ; but a particle cannot be moving in more than one direction at a time, so that what we have to consider is this :—as Newton virtually tells us that the presence of a second force in no way interferes with the action of the first, we have to seek first what are the effects of the two separately, and then what, in consequence of these effects supposed simultaneous, will be the actual motion of the particle. It comes then to be a question merely of compounding velocities—a purely geometrical (or, more strictly, *kinematical*) question instead of a physical one. The Second Law of Motion, therefore, enables us to commence with the purely kinematical notion of compounding two velocities, and thereafter to translate that into the compounding of two forces.

But the law of composition deserves a word or two. The compounding of two velocities is of course seen at once to be equivalent to this: If one body, such as a carriage, for instance, be moving in a certain direction with a certain velocity, and if some object in the carriage be simultaneously moving with reference to the carriage in a certain other direction, and with a certain other velocity, you can consider each of these separately— the motion of the carriage, or the motion of this body relatively to the carriage; but when you take the two simultaneously, the result is that, with reference to the ground supposed fixed, there is a perfectly definite direction and velocity with which the body is moving. This is an obvious truth; and the geometrical result is that, If we represent in magnitude and direction one of the two velocities by a line AB, and the second velocity by another line BC, drawn from the extremity of the first, then the single velocity, which is equivalent to the simultaneous existence of these two velocities, is found by drawing the third side AC of the hitherto uncompleted triangle. It follows then that (turning to the forces which produce these motions) as AB multiplied by the mass of the body is the change of motion produced by one of the forces, and BC multiplied by the same mass represents the change of motion produced by the second force,—the change of motion produced by the two forces acting simultaneously is the product of the mass moved into the third side AC of the triangle. But Newton's Law tells us that changes of motion are proportional to the forces which produce them. Therefore if AB be now taken to represent on a certain scale one of the forces, and BC the other, the single force which is represented on the same scale by the third side of the triangle will produce precisely the same effect upon the body as would be produced by the simultaneous action of the two separate forces. And you will see at once how it is that this law of geometrical composition of forces

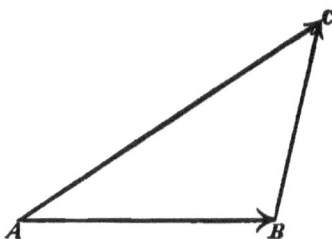

(what is called the triangle of forces), is merely a slightly different mode of expressing what you may be more familiar with under the designation of the parallelogram of forces, the so-called fundamental principle of statics.

There, then, is the law of the geometrical composition of forces, and also of velocities. We have in this case two sides of a triangle (taken consecutively and in the same way round), which may be said in a sense to be *geometrically* equivalent to the third side (taken the opposite way round), but the sum of their lengths is not equal to the length of the third side. This is the law of composition of what Sir W. R. Hamilton called *vectors*, and it is obviously generalisable into a similar construction for the composition of any number of velocities or forces in any directions in space. I leave it, without further comment for the moment, until I have made some remarks on Newton's Third Law, and then you will see how there is a physical sense in which we must take the sum, not of two sides themselves, but of the squares of two sides; how, in fact, the 47th proposition of the first book of Euclid comes in as part of the interpretation of Newton's Third Law of Motion.

Newton's Third Law of Motion, to which I have just referred, is expressed in very simple words: '*To every action there is always an equal and contrary reaction.*' These terms, 'action' and 'reaction,' Newton proceeds to explain. He tells us that there are two senses, quite different from one another, in which you may interpret each of these words; and yet that this same simple statement of the equality of action and reaction holds for each of these two perfectly distinct meanings.

The first form of action is that of an ordinary force or pressure; and Newton's statement then is equivalent simply to this: that if a weight presses upon a table, the table must react upon the weight with an equal and opposite pressure, and this whether the table is moving or not. Even supposing I were to lay so large a mass upon a table that the table were to give way, still while it was giving way, in the act of moving, if there were pressure at all between them, the load would press at every instant upon the table with an exactly equal and opposite force to that with which the table presses upon the load, and the same will hold however you may connect two bodies together. If you connect them either by mere contact, or by strings, or chains, or rods, or girders, anything—wherever there is a connection between two bodies—if there be any action whatever along that connecting link, there always is an equal and opposite reaction. And a visible or tangible link is not necessary. The same law is true of gravitation-attraction, and of electric and magnetic attractions.

So far, then, this is merely a question of forces; but it seems to have entirely escaped the notice, not only of Newton's contemporaries, but of those who have succeeded him during the last 150 or 200 years, until quite lately, that Newton's second explanation, his second mode of interpreting his Third Law, is something perfectly different from this, and leads us into a new order or range of phenomena. This second interpretation is so important that I must bestow considerable time upon it, because in reality it shows Newton to have been in possession of many of the principal facts of the conservation and transformation of energy. One or two of these facts escaped him, simply because he did not know what heat is, but he was very, very near attaining even that. He has given us all the mathematical materials that are required for the treatment of it; but he missed one great point, simply because experiment had not gone far enough in his time. Of course I need not say that he knew nothing (not even the name) of electro-magnetism and other recently discovered physical agents, all of which we can now classify under energy; but for everything that was known in his time, with the exception of heat, light, and electric energy, he gave us a complete statement. That complete statement, strange to say, has only been found in his great work within the last few years. It is this, literally translated :—'*If the activity of an agent be measured* [not by the agent itself, as in the case of a force, but] *by the product of its force into its velocity, and if similarly the counter-activity of the resistance be measured by the velocities of its several parts multiplied into their several forces, whether these arise from friction, cohesion, weight, or acceleration; activity and counter-activity in all combinations of machines will be equal and opposite.'* Now, in order to see the full force of this statement, let us consider what is meant by the product of a force into its velocity. Newton, as he has shown in a previous definition, understands, by the velocity of a force, not the whole velocity of the point to which it is applied, but the component of that velocity which is in the direction of the force. If, for instance, a horse is dragging a canal boat along, you are not to multiply the force of tension of the rope by the velocity of the canal boat, because the canal boat moves in one direction, and the tension of the rope is in general in a different direction. What you must do then is this: you must find out how much of the velocity of the boat is in the direction of the action of the force; resolve it, as it is called, multiplying the amount of the velocity of the boat by the cosine of the angle between its direction and the direction of the force which is applied by the rope. Then, what Newton says is this: if you so treat it—multiply each force by the velocity (in this sense) of its point of application—you will find that the sum of the activities will be equal to the sum of the counter-activities.

A word or two more about this before I consider the very admirable statement of various cases which Newton gives. Let us see what we mean now-a-days by what Newton calls here 'the action of the agent.' It is the product of the force into the resolved part of the velocity in the direction of the force. Therefore it is the product of the force into the rate at which the point of application moves in the direction of the force. The product of a force into the space through which it moves its point of application in its own direction is what we now call the amount of *Work done* by the force. But in Newton's statement it is not the amount of work done, but the rate at which work is being done, so that what he contemplates is really what we now-a-days measure, after Watt, by the unit called a horse-power—the rate at which an agent works when doing 33,000 foot-pounds of work per minute.

Now, you will particularly notice that he says the several parts of the resistance, 'whether these arise from friction, cohesion, weight, or acceleration.'

I shall take, first, cohesion and weight. You can easily see how a resistance may arise from cohesion, which simply means what we now call molecular forces in general, as, for instance, when work is spent in changing the shape of a body—when it is employed in producing a shear, for instance. There you have the elastic forces of the body worked against; and what Newton says is, that the amount of work spent, or the rate of spending work in distorting the body, is equal to the amount of work done or the rate of doing work against the elastic forces. It is thus stored up in the distorted body as Potential Energy.

Then he says 'weight :' the rate at which an agent works in lifting a mass is exactly equal to the rate at which work is done against gravity : and the work so done is stored up as Potential energy of the raised mass.

Then he says 'acceleration ;' and that is by far the most important of those I have yet mentioned. When work is spent on a body where there is no resistance from friction or from weight, or from cohesion, Newton says that work will always be spent against a resistance due to acceleration ; that is, work is spent in overcoming the inertia of a body and increasing its velocity. This is a statement of very great importance ; and when we interpret it according to Newton's previously laid down definitions, we find that his Third Law here asserts that the rate at which the agent works is the rate at which the kinetic energy of the body increases. For it is an immediate consequence of Newton's words that the rate at which work is spent is measured by the product of the momentum into the acceleration in the direction of motion. Hence the Kinetic Energy (which is half the product of the mass into the square of its velocity) is increased by an amount equal to the work spent. Work spent against resistance to acceleration is thus stored up in the body in the form of an increase in the kinetic energy.

That is very important; but there is a still more important point, which Newton takes account of, and that is, work spent against friction. Whenever work is spent against friction, we all know now-a-days that heat is produced, and it has been proved by elaborate experiments, which I shall presently discuss, that the amount of heat produced is precisely proportional to the amount of work spent in producing it. If Newton had known that such is the case, he could have had no difficulty whatever, after this extremely lucid statement of his, in passing to the general modern statement of the conservation of energy. So near had he arrived at it, that it wanted only experiments like those I am presently to describe, to have enabled him at once to

THE EARLY HISTORY OF ENERGY. 37

take a full grasp of the subject, at least so far as we know it in the present day.

Before I leave this matter, however, I must say a word or two as to the result of compounding two amounts of Kinetic energy. Suppose we have a southward velocity amounting, let us say, to 3 feet per second, and simultaneously an eastward velocity amounting to 4 feet per second, then we know by Kinematics, how to construct the single velocity, which is the resultant of these two. All we have to do is to draw a line of length 3 southwards, and from its extremity a line of length 4 to the eastward, and then complete the triangle. In a geometrical sense, therefore, a velocity of 3 southwards and a velocity of 4 eastwards will be equivalent to a velocity which, if you calculate what the third side of that triangle will be, is represented by 5 on the same scale. It will then be a velocity of 5 in a direction which makes an angle, whose sine is $\frac{4}{5}$, with the south line. So far the geometrical conception of composition is perfectly definite. But now let us see what this involves in the case of Kinetic energy. If a mass were moving with a velocity of 3 southwards, and simultaneously with a velocity of 4 eastwards: its Kinetic energy, being proportional to the square of the velocity, is in the southward direction proportional to 9, the square of 3, while in the eastward direction it is proportional to 16. But the same mass moving with the resultant of these velocities has Kinetic energy proportional (on the same scale) to 25—the *arithmetical* sum of the other two. So that there are two ways of compounding these combinations of the velocity and mass of a body. When it is a question of Momenta

—that is to say, when it is a question of the application of Newton's first meaning of the word *actio*—when the *actio* means a simple force or its time-integral, then you are to compound geometrically, and two sides of a triangle are in that sense equal to the third; but when it comes to compounding Kinetic energies which are proportional to the square of the velocity, then you are limited to right-angled triangles, and having to add the squares of the two sides, you obtain the square of the third side. The difference then between the geometrical composition and the simple arithmetical addition is a difference depending upon the use of the first or second power of the velocity. When, as in momentum, the first power is involved, the magnitude is essentially a directed one, and two directed magnitudes must be compounded geometrically. But Kinetic energy, depending as it does upon the square of the velocity, is essentially non-directional, and its various parts, when independent of one another (as they are when they depend upon motions in directions perpendicular to one another), are to be compounded by simple addition. These two things, then, Momentum and Kinetic Energy, perfectly distinct from one another, having no reference to one another that we can trace at present, are both included in the simple form of statement of Newton's Third Law, only with a corresponding difference of meaning to be attached to two of the words involved.

What Newton really wanted then was to know what becomes of work which is spent in friction. Now, the first successful answerer of that question was undoubtedly Count Rumford, and from his paper of 1798 I shall read some extracts, because it is one of the most valuable experimental papers that perhaps ever was published. It is most admirably philosophical in its mode of experimenting, and it is throughout entirely opposed to that *à priori* style of reasoning which (as I showed you in my last lecture) is so fatal to progress in natural philosophy. Count Rumford says :—

'It frequently happens, that in the ordinary affairs and occupations of life, opportunities present themselves of contemplating some of the most curious operations of Nature ; and very interesting philosophical experiments might often be made, almost without trouble or expense, by means of machinery contrived for the mere mechanical purposes of the arts and manufactures.

'I have frequently had occasion to make this observation ; and am persuaded, that a habit of keeping the eyes open to everything that is going on in the ordinary course of the business of life has oftener led, as it were by accident, or in the playful excursions of the imagination, put into action by contemplating the most common appearances, to useful doubts, and sensible schemes for investigation and improvement, than all the more intense meditations of philosophers, in the hours expressly set apart for study.'

Then again he says :—

'Being engaged, lately, in superintending the boring of cannon, in the workshops of the military arsenal at Munich, I was struck with the very considerable degree of Heat which a brass gun acquires, in a short time, in being bored ; and with the still more intense Heat (much greater than that of boiling water, as I found by experiment) of the metallic chips separated from it by the borer.

'The more I meditated on these phenomena, the more they appeared to me to be curious and interesting. A thorough investigation of them seemed even to bid fair to give a further insight into the hidden nature of Heat ; and to enable us to form some reasonable conjectures respecting the existence, or non-existence, of an *igneous fluid :* a subject on which the opinions of philosophers have, in all ages, been much divided.

'From *whence comes* the Heat actually produced in the mechanical operation above mentioned ?

'Is it furnished by the metallic chips which are separated by the borer from the solid mass of metal ?

'If this were the case, then, according to the modern doctrines of latent Heat, and of caloric, the *capacity for Heat* of the parts of the metal, so reduced to chips, ought not only to be changed, but the change undergone by them should be sufficiently great to account for *all* the Heat produced.'

He sees the difficulty : he catches at once really what is wanted—the true method of upsetting the old notion that heat is matter. The explanation which was given of the heat produced by friction by those who believed that heat is matter was simply this:—The body in its solid state, or rather in its massive state, before you began to abrade filings from it, possessed in that state a certain quantity of heat. It had a certain capacity for heat at a certain temperature ; in other words, it required so much heat to be mixed up with its particles in order to make the temperature of the whole that which was observed. But if you could make it more capacious—if you could give it greater capacity for heat—then it would hold more heat without becoming of a higher temperature. On the other hand, if by any process whatever you could diminish its capacity for heat, then, of course, it would become hotter itself, and even give out heat to surrounding bodies, so that, according to the notion of the supporters of the caloric theory (as it was called), the production of heat by friction or abrasion is due to the fact that you make the capacity of a body for heat smaller by reducing it to powder. For of course, when its capacity for heat is thus made smaller, it must part with some of the heat it had at first ; or if it retains it, it must necessarily show the effect of the heat more than it did before, and must therefore rise in temperature. Now, this reasoning is, so far, perfectly philosophical.

THE EARLY HISTORY OF ENERGY. 41

We can say nothing against a mode of reasoning of that kind. The only fallacy in it was the assumption that heat is a substance. Now, see how well Rumford laid hold of that point, and how he proceeds by experiment to try if possible to satisfy his doubts about it. He says :—

'If this were the case, then, according to the modern doctrines of latent Heat, and of caloric, the *capacity for Heat* of the parts of the metal so reduced to chips, ought not only to be changed, but the change undergone by them should be sufficiently great to account for *all* the Heat produced.'

Rumford found no difference, so far as his form of experiment enabled him to test it, between the capacity for heat of the abraded metal and the metal before the abrasion had taken place; so that if this experiment had been only a satisfactory one—and Rumford did not see how to make it thoroughly satisfactory—the fact that heat is not matter would have been conclusively established. What Rumford really did want was this: he wanted a process by which to bring the abraded metal and the non-abraded metal, if possible, to the same final state. He tried to do this by throwing them into water—equal quantities of the lumps and of the filings, equally hot, into equal quantities of water at the same lower temperature—to see whether they would produce different changes of temperature, each in its own vessel of water. But then they were not in the same final state. The filings, remember, were in a distorted state; they might have been very considerably compressed, or they might have been distorted in shape by shearing or something of that kind, in virtue of which they might have had a certain quantity of latent heat which he could not discover by this process. The only legitimate and practicable process which we know of for completely answering that question, which was Rumford's sole difficulty, is a chemical process. Dissolve your lumps and an equal weight of your filings in equal quantities of the same acid. At the end of the operation, of course, there can be no doubt that the chemical substances produced will be precisely the same, whether you begin with lumps or with filings. You will have the same chemical substance; but if there be any mysterious difference as to the capacity for heat in them, that will be shown during the process of solution. In general, in dissolving a metal in an acid, there is a development of heat; but if there were any difference in the quantity of heat which the lumps and an equal weight of filings contained—that is to say, if heat could by any possibility be matter—then there would necessarily have been an escape of heat more in one vessel than the other. If Rumford had tried that one additional experiment, he would have had the sole credit of having established the non-materiality of heat.

The details of Rumford's experiments are given in full, but I shall not describe them to you. I merely mention that they show extraordinary skill and care in experimenting, and wonderful precaution in trying to avoid, as far as possible, the necessary losses in the experiments. When losses were unavoidable and of a large amount, the same skill is shown in making separate side experiments, in order to enable the operator to allow for them in the main experiments. The whole work itself is a model of experimental science. I shall now pass on to the final reasoning, merely mentioning in passing that Rumford actually managed to boil a large quantity of water, though an immense

THE EARLY HISTORY OF ENERGY. 43

amount of heat was lost in spite of all his precautions. Still the work of a single horse for two hours and twenty minutes was found sufficient to boil about 19 lbs. of water, besides heating a large casting of the cannon, and all the machinery that was engaged in the process. He says :—

'It would be difficult to describe the surprise and astonishment expressed in the countenances of the by-standers, on seeing so large a quantity of cold water heated, and actually made to boil, without any fire.

'Though there was, in fact, nothing that could justly be considered as surprising in this event, yet I acknowledge fairly that it afforded me a degree of childish pleasure which, were I ambitious of the reputation of a *grave philosopher*, I ought most certainly rather to hide than to discover.'

Here is his final reasoning :—

'In reasoning on this subject, we must not forget to consider that most remarkable circumstance, that the source of the Heat generated by friction in these experiments, appeared evidently to be *inexhaustible*.

'It is hardly necessary to add, that anything which any *insulated* body or system of bodies can continue to furnish *without limitation*, cannot possibly be *a material substance*. It appears to me to be extremely difficult, if not quite impossible, to form any distinct idea of anything capable of being excited and communicated in the manner in which the heat was excited and communicated in these experiments, except it be motion. I am very far from pretending to know how or by what means or mechanical contrivance that particular kind of motion in bodies which has been supposed to constitute Heat is excited, continued, and propagated;'

and then he proceeds to apologise for the minutiæ given in his paper.

Now, when we make a calculation from the data furnished by Rumford's paper, we find this : that, supposing heat to be a form of energy, and taking 30,000 foot-pounds per minute as the work of a horse (that is something like an ordinary estimate), the mechanical equivalent of heat is 940 foot-pounds. The meaning of that statement is, that if you were to expend the amount of work designated as 940 foot-pounds in stirring a single pound of water, then that pound of water when brought to rest at the end of the operation would be one degree Fahrenheit hotter than before you commenced. [Rumford throughout uses Fahrenheit's degrees.] We can put it in another form, which is perhaps still more striking. If you had a cascade or waterfall 940 feet high, then, in the fall of the water down that cascade, there would be 940 foot-pounds of work done by gravity upon each pound of water; and therefore if all the energy which the moving water has, as it reaches the bottom of the fall, were spent simply in heating the water, the result would be that the water in the pool at the bottom of the fall would be 1 deg. Fahrenheit hotter than the water at the top of the fall.

I may remind you here, that Rumford's experiments were published in 1798, so that they are of considerably old date; but, like those which I am just going to advert to, they were barely noticed, or noticed only to be laughed at, until somewhere about the year 1840.

Now, in the very year after the experiments of Rumford were published, we had the experiments of Davy. I need not go into minute details about them, because they were not by any means such models of careful experimental work as Rumford's. But, for all that, Davy gave conclusive proof (if he had only at the time seen it himself) that heat is not matter. His proofs were of this kind. He first showed that by rubbing two pieces of ice together—by simply expending work in the friction of two pieces of ice—you could melt the ice. Now, supposing heat had been matter, this is the sort of argument that a believer in the caloric theory would have used: two pieces of ice when rubbed together cannot possibly melt one another, because in order to melt them you will have to furnish heat to them. But the heat can only come from themselves when they are rubbed together; it cannot come from surrounding bodies, and therefore they cannot possibly melt together, because to melt one another, they would have first to part with some of their heat in order to produce the melting. Davy showed, however, that the mere rubbing together of two pieces of ice by proper mechanical processes was sufficient to melt the surface layer of each. There still was this possible objection, that the heat might have come from some external source, so that his second experiment was of this kind. He rubbed two pieces of metal together, keeping them surrounded by ice, and in the exhausted receiver of an air-pump, so as if possible to avoid radiant heat, heat carried by convection-currents of air, and so on, and to remove every possible disturbing cause, or even source of suspicion, from his experiment; and still he found that these two pieces of metal, when rubbed together thus, constantly produced heat and melted the ice, every precaution having been taken to prevent heat from getting at them from every side. It is curious that his reasoning upon the subject is extremely inconclusive, although his experiments themselves completely settle the question. He says:—

'From this experiment it is evident that ice by friction is converted into water, and according to the supposition its capacity is diminished; but it is a well-known fact that the capacity of water for heat is much greater than that of ice ; and ice must have an absolute quantity of heat added to it before it can be converted into water. Friction, consequently, does not diminish the capacities of bodies for heat ;'

and there he stops. [Sir W. Thomson remarks on this passage (*Encyc. Brit.*, last edition, art. *Heat*), as follows :—Delete from 'and according to the supposition,' to 'greater than that of ice,' inclusive ; and delete the lame and impotent conclusion stated in the last eleven words. The residue constitutes an unanswerable demonstration of Davy's negative proposition that heat is not matter.] But some years afterwards he came to this conclusion from these experiments :—

'Heat, then, or that power which prevents the actual contact of the corpuscles of bodies, and which is the cause of our own sensations of heat and cold, may be defined as a peculiar motion, probably a vibration of the corpuscles of bodies tending to separate them. It may with propriety be called the repulsive motion. Bodies exist in different states, and these states depend upon the action of attraction and of the repulsive power on their corpuscles, or, in other words, on their different quantities of repulsion and attraction.'

Now, we see at a glance how he explains by these experiments what is the difference between a solid and a liquid, and the difference again between a liquid and a gas. In general, the melting of a solid is produced by communicating heat to it. In other words, according to Davy's explanation, the particles of the solid are set in vibration, and thus, in consequence of the repeated impacts upon one another, they push one another side. And, as he also says, you may consider this repulsive motion to have a complete analogy to the so-called centrifugal force in a planetary orbit, for the faster one particle is moving about another, the larger necessarily is the orbit into which it will be forced. The particles of a solid then are forced from one another by this repulsive action of heat, and the action of the heat upon it puts it into a liquid state. When you increase still further the amount of heat communicated to the body, you at length overcome altogether the cohesive forces, and you have free particles, as in a gas, flying about and impinging upon one another, but only for very brief periods coming near enough in the course of their gyrations to bring into play the molecular forces again. Whenever, however, the molecular forces do come into play for a moment, you may have two particles adhering together, but they are soon knocked asunder by a blow from a third particle.

There is one other sentence, however, which I must quote from Davy, and then I shall have finished my account of his contributions, which were later than 1799, when his first paper was published. In fact, in 1812 he enounces this proposition :—

'The immediate cause of the phenomenon of heat, then, is motion, and the laws of its communication are precisely the same as the laws of the communication of motion.'

Now, we see at a glance to what an immense extent the science had been advanced in Davy's time. When Davy was in a position to make that statement, one had only to take it in addition to the second interpretation of Newton's Third Law, and the dynamical theory of heat was in his possession. Still, that publication of Davy's in 1812, like the earlier ones of Rumford and of Davy himself, remained almost unnoticed—looked upon, perhaps, as an ingenious guess, or something of that sort, but as something which it was not worth the trouble of philosophers to consider; and it was not until Joule's time, somewhere about 1840, that the subject was fairly taken up, and that justice was rendered to their real value. Notice how distinctly these two great leaders were men who based their work directly upon experiment. There is no *à priori* guessing, or anything of that kind, about either Rumford's or Davy's work. They simply set to work to find out what heat is. They did not speculate on what it might be. But both before and after their time, there have been numbers of philosophers who have, without trying a single experiment, or at best trying only the roughest forms of experiment, endeavoured to discover by *à priori* reasoning what heat is. The list is a very long one, and includes names such as Locke and Bacon, which are distinguished in very different subjects, as well as to some extent in physics. These both express their complete conviction that heat consists in a brisk agitation of the particles of matter; but then, as this was based upon no experiment whatever, it can simply be looked upon as a happy guess. In the present day when a philosopher comes forward and makes similar statements, without any experiment, we simply put him in the same category as Locke and Bacon, we justly refuse to give him any credit for a matter of that kind.

There was one man of this class, however, M. Séguin, a nephew of the celebrated Montgolfier, who all but redeemed himself from being so classified. Séguin himself says he got from his uncle his idea that heat is certainly not matter, but corresponds to a certain kind of energy; and he says that he had made various experiments with a steam-engine, in order to test whether the same quantity of heat reached the condenser as had left the boiler. He was, unfortunately, unsuccessful in all his experiments. He was certainly on the right track, and had he succeeded there he would have been entitled to be considered as an independent discoverer of the non-materiality of heat. For it is obvious that if we can show by any experiment whatever that heat is put out of existence, or that fresh heat is brought into existence, either of these at once destroys all possibility of its being material. Now, if Séguin could have proved, by his actual measurements, that less heat in any case reaches the condenser than left the boiler, he would have completely settled the question. From that point of view experiments have been made, and made very carefully, in recent times, by Hirn. Hirn has actually measured, in an ordinary working steam-engine, by most careful experimental methods, the quantity of heat which leaves the boiler and the quantity which reaches the condenser. He has measured also the quantity which is lost by radiation, conduction, and currents of air over all parts of the machine, and he has found, as a final result, that when the engine is at work, as, for instance, when a number of spindles are being turned, there is a greater difference between the quantity of heat which leaves the boiler and the quantity which reaches the condenser than when the steam is simply blown through the engine without doing any work. In the latter case the greater part of it reaches the condenser; in the former case there was less of it that reached the condenser—more of it, in fact, was put out of existence, or, to speak more correctly, more of it was converted into work done by the engine during the operation.

But what I chiefly wish to impress upon you is that

Séguin, although he went to work in a correct manner, reasoned from an utterly unsound basis. His reasoning was of this kind, that when a body expands and thereby becomes colder, it loses heat, and that the heat so lost is necessarily the equivalent of the work done during the expansion.

Another of the speculators on the dynamical theory of heat, but who did not publish till 1842, three years after Séguin, was Mayer, who in very many quarters still gets the credit of being the real author of the whole science of Energy, including Thermo-dynamics. Mayer's speculation was based on precisely the converse of that of Séguin. Séguin said the amount of work done by an expanding heated body is the equivalent of the heat which it loses. Mayer said the amount of heat which is produced by compressing a gas or any other body is the equivalent of the work spent in compressing it. You will see at once that these two statements are precisely the same, only the one is the converse of the other. If the one be true, the other necessarily will be true also; but both are *à priori* assumptions, and we now know by experiment that neither of them is true under any realisable circumstances whatever; though in certain cases they are approximately true. Each of the two speculators, Séguin and Mayer, tried to apply his hypothesis by calculation to the properties of a particular substance. Séguin tried steam, because he was more familiar with steam; Mayer tried air, because he had some physical data for it. Séguin's calculations were very far wrong on the one side of the truth; Mayer's were very far wrong on the other side of the truth; but Mayer's substance, namely, air, has been since experimentally proved by Joule to be capable of giving an almost exact result. Mayer by chance, then, in the middle of his *à priori* speculations, lit upon a method—although he got it from a false principle—which Joule afterwards proved to be a good one, and used as one of his modes of obtaining the value of the dynamical equivalent of heat. Still, we must give Mayer no credit for that, for although he laid down his law quite generally, air was the only substance he had data for, and he chose it on that account. But even with this, his data were so bad that he got a result as far from the truth as the one obtained by Séguin. Only Séguin has this great credit, to which Mayer has no claim, that, seeing that if heat be not matter, some of it must disappear in the working of an engine, he tried to measure the quantity of heat coming to the condenser, in order to show that it was less than that which left the boiler.

I find that I have now exhausted my time, and therefore I shall merely mention that, in my next lecture, I shall take up the history of the theory of energy, as it was developed by the sound methods of Colding and Joule in papers published about 1843; and I shall then endeavour, with the facilities which this room affords me, to illustrate my explanation by a few experiments.

[*Note to Third Edition.* The last three or four pages have been left in their original form, as expressing what was well known in 1874. But, of late, attention has been called to the services of Mohr, whose date is prior to that of Séguin, and still more so to that of Mayer. In the next lecture, a notice of these services will be inserted.]

LECTURE III.

ESTABLISHMENT OF THE CONSERVATION OF ENERGY.

Further inquiry into the asserted claims of Mayer. Opinions of Colding and Joule on Mayer's first paper. [Insertion (1884) on the prior claims of Mohr.] Colding's Experiments. Joule's Experiments. Numerical value of the Dynamical Equivalent of Heat. Helmholtz's argument from the Perpetual Motion. Transformation and Dissipation of Energy. Illustrative experiments.

IN my last lecture I showed you in what state Newton left the grand question of conservation of energy, what an enormous step he took, and what was the sole great difficulty remaining in his way. Then I showed you how, in regard to the particular branch of it which we call the dynamical theory of heat, Rumford and Davy had, at the very end of last century, almost completely settled the question that heat is not matter. A little was wanting in the work of each. Rumford wanted only one small chemical experiment in addition to his grand physical experiments. Davy wanted a little more conclusive reasoning than he showed at the time. Had one or other of these been furnished before the end of the last century, it would have been to the last century that we should have been indebted entirely for the dynamical theory of heat. It was not, however, until 1812 that Davy applied correct reasoning to his experiments, and obtained the correct deductions from them; and then he stated in a distinct form the important propositions that heat is motion, and that the laws of its communication are precisely the same as the laws of communication of motion. Then I showed you that Séguin, although he was altogether wrong in his *à priori* idea, had a true sense of what was really wanted to this question, and that he made a correct, but unhappily unsuccessful, experimental attempt to supply it. Then we came to Mayer, a man who has, especially of late, been persistently held up as the discoverer, not merely of the dynamical theory of heat, but of the whole subject of conservation of energy. Of him, I may remark —because the question is one of importance—though at the present day we are hardly perhaps far enough advanced in time calmly and dispassionately to consider the relative claims of these authors; still, I may remark that a great deal of the eulogy which has been bestowed upon Mayer is altogether undeserved, and that Joule has even yet received far too little credit for the enormous advances he made. In the first place, Mayer was altogether wrong in his *à priori* idea. On that Sir William Thomson and I made, in 1862, the following remarks, which no one has ventured directly to challenge in the slightest particular :—

'Mayer's method is founded on the supposition that diminution of the volume of a body implies an evolution or generation of heat; and it involves essentially a false analogy between the natural fall of a body to the earth, and the condensation produced in an elastic fluid by the application of external force. The hypothesis on which he thus grounds a definite numerical estimate of the relation between the agencies here involved, is that the heat evolved when an elastic fluid is compressed and kept cool, is simply the dynamical equivalent of the work employed in compressing it. The experimental investigations of subsequent naturalists have shown that this hypothesis is altogether false for the generality of fluids, especially liquids, and is at best only *approximately* true for air; whereas Mayer's statements imply its indiscriminate application to all bodies in nature, whether gaseous, liquid, or solid, and show no reason for choosing air for the application of the supposed principle to calculation; but that at the time he wrote, air was the only body for which the requisite numerical data were known with any approximation to accuracy.'

Then, in addition to these two absolute errors which are mentioned in this passage, I may call attention to the preposterous *à priori* principles upon which he reasons. There are two of them; the one is *causa æquat effectum*, to which I have never been able to attach any meaning, and the other *ex nihilo nihil fit*. These may be a basis for scholastic disquisitions, such as the celebrated old question of the number of angels that can simultaneously dance on the point of a needle, but they are altogether unfit for introduction in any shape whatever into physical reasoning. Then, again, Mayer's work was altogether destitute of experiment. He suggests, no doubt, the carrying out, on a larger scale, an experiment which he says he tried, namely, shaking a little phial of water for a considerable time, to find it at the end of the time warmer than it was at the commencement:—merely, I may say in passing, a bad substitute for a hint due to Rumford, that the *churning* of water would be a good experimental method. I daresay most of you will see that such an experiment as Mayer's, unless proper precautions were taken to prevent conduction of heat from the hand to the bottle of water, would very probably have resulted in the heating of the water considerably, even without the shaking: so that, in order to prove that the heat was due to the shaking, we should have required at all events a statement on Mayer's part of the precautions he had taken to prevent one known source of heat from affecting the water.[1]

But, in addition to this, Mayer did not even believe that heat depends on motion ; and this is perhaps the most wonderful comment that can be made upon the consistency of those who, while constantly speaking of heat as a 'mode of motion,' call him the discoverer of the modern theory of heat. To effect this must surely have involved (to use the vigorous and expressive language of one of the most prominent popularisers of science) the necessity of 'wrangling resolutely with the facts!' Mayer himself says, in his very earliest paper, and he never afterwards to my knowledge modified this statement (I translate freely), 'We might much rather assert the opposite, that motion, whether it be a simple one or a vibratory one—like light, like radiant heat, and so on—must, in order to become heat, cease to be motion.' He actually says it must *cease to be motion* in order to become heat! Then he makes another and a very curious statement, the absolute erroneousness of which you will see in the course of another lecture. He says, sneeringly: 'Let any one try to melt ice by pressure, however enormous.' I shall show you that, as a consequence of the second law of thermo-dynamics, the melting of ice by pressure was predicted beforehand, and was verified afterwards by actual experiment.

It is time, then, I say, that Mayer, even with our as yet imperfect means of judging, should be ranged, so far as we can, in his true place. He has been injudiciously praised, and he has been an unfortunate man,

---

[1] Even this experiment, but carried out with something like philosophical precautions, was long before described, by Reade in *Nicholson's Journal*, 1808, p. 113.

and therefore, of course, there will be an outcry against any one who undertakes the necessary task of pointing out his real demerits. However, there is no such thing in scientific history as the *argumentum ad misericordiam*. The blame, if any there be in such a matter, is due to those who preposterously gave him credit for what he did not do. The real merits of Mayer, however, which are extremely great, but which are in danger of being forgotten or ignored in consequence of the unwarrantable claims made for him, depend upon his having, after getting a true theory by false reasoning from inadequate and sometimes inadmissible premises, reasoned rightly upon it, and developed it widely in its applications. Language has lost all meaning, however, if this can be called a claim to establishment of the theory itself. The fact is that in 1839 Faraday, and in 1841 Liebig, and about the same time others of the great philosophers who have lately died, made close approaches to the true theory by methods far more sound than those of either Mayer or Séguin ; and yet, curiously enough, they have scarcely at any hand got the slightest recognition.[1]

The true modern originators and experimental demonstrators of the conservation of energy in its generality were undoubtedly Colding of Copenhagen and Joule of Manchester. It is interesting to see in what light these men regard Mayer and some others of those who preceded them. I shall presently give you a quotation or two bearing on that point.

In the meantime I may say, with regard to Colding,[2] that he began by being metaphysical, but saw at once,

[1] See *Phil. Mag.* 1864, II. p. 474; 1865, I. p. 217 ; and 1876, II. p. 110.
[2] See his very interesting letter, *Phil. Mag.* Jan. 1864.

or very soon, that metaphysics was not the proper basis on which to found a search for physical facts. His metaphysics led him to form certain opinions, but before publishing one of them he set to work and laboriously brought it to the test of fact. Joule, on the other hand, seems to have begun by experimenting with the view of determining certain physical constants. He does not tell us whether he had any metaphysical opinion about their relations or not. He set to work experimenting, and it was only after a great and varied series of his experiments had been fully carried out, and valuable results obtained, that he began to make certain applications of metaphysical reasoning to the connections which he had discovered. He did not apply metaphysics to discover anything, but to try and co-ordinate with other things the discoveries he had already made. Colding's work is by no means so extensive as Joule's. It is very nearly simultaneous with it, but it is neither so exact nor so extensive. Still, although Colding is hardly to be compared with Joule, he stands enormously high in comparison with any of the others who had experimented up to that time upon the conservation of energy. I will read you one or two extracts from Colding, and you will see from them how properly he went to work. He says :—

'It was in accordance with this idea that I twenty years ago presented to the Royal Society of Science here in Copenhagen, a treatise in which I explained my idea that force is imperishable and immortal ; and, therefore, when and wherever force seems to vanish in performing certain mechanical, chemical, or other work, the force then merely undergoes a transformation and reappears in a new form, but of the original amount as an active force.

'In the year 1843 this idea, which completely constitutes the new principle of the perpetuity of energy, was distinctly given by me, the idea itself having been clear to my own mind nearly four years before, when it arose at once in my mind by studying *D'Alembert's celebrated and successful enunciation of the principle of active and lost forces;* but of course the new principle was not as clear to me from the beginning as it was when I wrote my treatise in 1843.'

I may here parenthetically observe that Colding speaks of D'Alembert's celebrated and successful enunciation of a certain principle. This is nothing more or less than a particular case of that principle of Newton, which I gave you in a former lecture;[1] so that you see Colding really got his idea suggested to him by Newton's work :—

'According to the view which led me to this principle, its future importance, in case it were really true, was perfectly clear to me from the first instant. But this made me very anxious not to publish it as a new law of nature until I should be able to give experimental proof of its truth ; and scientific men to whom I explained my idea, and especially our celebrated professor, H. C. Œrsted, agreed with me and advised me to be safe in this respect before I wrote ; and it was for this reason that I departed from my original intention of explaining it to a meeting of Natural Philosophers held in Copenhagen in 1840.

'In my first treatise, of 1843, the title of which is " Theses concerning Force" (*Nogle Sætninger om Kræfterne*), I therefore not only presented my idea to the Royal Society (of Copenhagen) as a thing that most likely would hereafter be found to be a general law of nature, but, after stating that the only trustworthy decision of the question was to be got from the experimental investigation of nature itself, I went on to call attention to several old experiments made previously to my time, the first of which was Dulong's celebrated discovery respecting the heat disengaged or absorbed during the compression or expansion of a great number of different airs and gases, and I then showed how perfectly these experiments proved the truth of the said principle for bodies of that kind.'

[1] *Ante,* p. 33. See Thomson and Tait's *Natural Philosophy,* § 264.

Then he goes on to say that having established the proposition for elastic fluids, he proceeded to try experiments in conjunction with Œrsted upon the compression of water; and that next he advanced, just as Joule did about the same time, to experiments upon the compression of solids. He also says:—

'I closed my discussion by showing that the discovery of a *perpetuum mobile* would be possible if my principle was wrong.'

This shows that, to a certain extent at least, he had anticipated Helmholtz, of whose great services to this branch of science I shall presently speak.

The remarks he makes about Mayer deserve to be quoted. He desires the republication, in an English journal, of his first paper, in order that it might be compared, as he says, with the paper of Mayer, which was most loudly vaunted in England at the time when his letter was written:—

'I need scarcely say that such a comparison would be of great interest to me, as I believe it would convince your readers of the fact that M. Mayer wrote his remarks in 1842, before he was able to support them by a single experiment or by anything like a proof of their exactness, whilst I thought it to be my duty, before I wrote, to prove that my suppositions concerning the forces were confirmed by nature itself, as a law of nature.'

He also says of his own *experimental* approximation to the dynamical equivalent of heat, that it is

'very near the proportion that M. Mayer in 1842 *supposed*, but *did not prove*, to be right.'

Joule's remarks[1] upon the subject of Séguin and Mayer are also deserving of quotation:—

'Séguin gives data from which the mechanical equivalent of heat may be readily deduced on his hypothesis, the result being too

[1] *Phil. Mag.* 1864, II. p. 151; see also 1862, II. p. 121.

great in consequence of the thermal effect of the compression of vapour being understated. Neither in Séguin's writings of 1839, nor in Mayer's paper of 1842, were there such proofs of the hypothesis advanced as were sufficient to cause it to be admitted into science without further inquiry. I believe that the experiment attributed to Gay-Lussac was not referred to by Mayer previously to the year 1845. Mayer appears to have hastened to publish his views for the express purpose of securing priority. He did not wait until he had the opportunity of supporting them by facts. My course, on the contrary, was to publish only such theories as I had established by experiments calculated to commend them to the scientific public, being well convinced of the truth of Sir J. Herschel's remark, that "hasty generalisation is the bane of science."'

To these it would be easy to add several even more telling passages to the same effect.

[In 1876 my attention was called to a paper by Mohr (*Journal für Pharmacie*), of which I published a translation in the *Phil. Mag.* for August of that year. The date of the paper is 1837, or five years before Mayer, and it contains, in a considerably superior form, almost all that is correct in Mayer's paper. Though it contains many mistakes, it avoids some of the worst errors of Mayer, especially his false analogy and his *à priori* reasoning. The very process (for determining the mechanical equivalent of heat from the two specific heats of air) for which Mayer has been so extravagantly lauded:—although it is in principle, albeit not in practice, utterly erroneous :—is here stated much more clearly than it was stated five years later by Mayer.

In December 1877, I received by post a copy of a work *Allgemeine Theorie der Bewegung und Kraft*, etc. (Braunschweig 1869), with the inscription, in a bold hand, 'dedicated by the author, Dr. Mohr.' This work is conclusive against Mayer's first paper. It leaves absolutely nothing to him save his blunders. For it contains a reprint of an article by Mohr, published in 1837 in *Baumgartner's und v. Holger's Zeitschrift für Physik* (of which the paper above alluded to was, it seems, a mere *résumé*). One sentence, only, need be extracted from this article (which ought certainly to be translated into English *verbatim*) to show how definitely in 1837 Mohr put into words a clear statement of the truth which Mayer vainly attempted to express clearly five years later.

'Ausser den bekannten 54 chemischen Elementen gibt es in der Natur der Dinge nur noch ein Agens, und dieses heisst *Kraft:*—es kann unter den passenden Verhältnissen als Bewegung, chemische Affinität, Cohäsion, Electricität, Licht, Wärme und Magnetismus hervortreten, und aus jeder dieser Erscheinungsarten können alle übrigen hervorgebracht werden. Dieselbe Kraft, welche den Hammer hebt, kann, wenn sie anders angewendet wird, jede der übrigen Erscheinungen hervorbringen.'

This notable article did not obtain insertion in *Poggendorff's Annalen*, to which it was first sent. One of the earliest and most valuable of Joule's papers met a similar fate at the hands of the *Royal Society*.]

Having said this much with regard to the relative merits of these men, and having shown you that Joule is far the foremost, while Colding is the only one who deserves mention in comparison with him, so far as the present part of our subject is concerned, I proceed to give a rough general statement of what Joule really did, and then you will see what enormous advances he made within a few years from 1840. Joule, in 1840, published his first paper, which was with reference to the heat produced by electric currents under various circumstances. He was led by these experiments to see that there must be some relation between the heat produced and the quantity of zinc consumed in the battery; thus, as it were, eliminating the mysterious agent, electricity, altogether from the final result. The novelty and value of this idea can hardly now be realised by us. Then, again, Faraday's grand discovery of induced currents suggested to Joule the measurement of the amount of mechanical work we require to spend in order to produce a given amount of electric current, which in its turn shall be frittered down into a given amount of heat. We should thereby have, as it were, not an immediate conversion of work into heat, as in the case of friction (which appears at least at first sight to give an immediate transformation from work into heat), but we should have a mediate transformation by induction of currents —we should transform the work of driving the magneto-electric machine into the energy of so much electric current, and then let that again turn itself into heat. You have first the work, then the electric currents, and finally the heat. Now, Joule seems to have observed that the same amount of heat was produced from this amount of work, whether the work was first employed in producing electricity, and then the electricity employed in producing heat, or whether the work was simply spent directly in producing heat by friction; and from that time he began to experiment, with the view of determining exactly what is the mechanical equivalent of heat, because he saw that unless it were certain, experimentally, that in all cases of friction, where there is nothing but heat to show for the work that has been spent—unless there could always be found the same amount of heat for the same amount of work, whatever were the bodies which were made to rub against each other—unless something of that kind could be established, it would be vain to seek for any such thing as conservation of energy, or even for the much lower and in fact mere particular case of the equivalence between heat and work. If work and heat be equivalent in any sense, and if you spend work wholly in producing heat, you must get always the same amount of heat for the same amount of work, whatever be the nature of the engine which you employ. I may parenthetically remark (as it gives an inkling of what is to follow) that it is quite another question when you come to the conversion of heat into work ; when it comes to be a question of beginning with the heat, and converting that into work, the conversion cannot be wholly accomplished. Begin with work, and you can convert it all into heat. Begin with heat, and you cannot convert it all into work. The one case is perfectly definite, and therefore Joule, reasoning upon it, virtually said :—' If there be nothing but heat to show for a certain amount of work spent, then unless we always get, with every apparatus, the same amount of heat for the same amount of work, conservation cannot possibly hold.' He proved that this equivalence does subsist ; and his determination, finally published with all his latest improvements in 1849, was 772 foot-pounds for a unit of heat ; that is to say, a pound of water which has fallen 772 feet, and had the whole of the energy of its fall, or the whole excess of potential energy which it had before falling, converted into heat, will simply be 1 deg. Fahr. hotter than it was before it fell. As I pointed out to you in my last lecture, Rumford's estimate was considerably above that ;

but it was confessedly only an estimate, while Joule's was the final result of an extended and laborious series of experiments. This leads us then to the statement of what is called the *First Law of Thermo-dynamics*. It may be put in very many forms, but I shall take the form which seems to be the most effective. The first law of thermo-dynamics, then, really established by Davy and Rumford, but altogether neglected and forgotten, re-established by Joule and supplied by him with a definite numerical *datum*, for the purpose of calculation, may be put in this form :—

When equal quantities of mechanical effect are produced by any means whatever, from purely thermal sources, or lost in purely thermal effects, then equal quantities of heat are put out of existence or are generated: and for every unit of heat measured by the raising of a pound of water 1 deg. Fahr. in temperature, you have to expend 772 foot-pounds of work.

It is possible that that last figure of the 772, which is for the latitude of Manchester, may be wrong. The true number may be, for instance, 771·5 or 772·5, or something of that kind, but there is little doubt that Joule's determination is at all events considerably within one per cent. of the truth. It is particularly noteworthy that in 1843, from the heat developed by the friction of water in narrow tubes, Joule had given 770 foot-pounds as the mechanical equivalent.[1]

In addition to all this, Joule gave an experimental extension of the principle of conservation to other forms of energy,—that is to say, in addition to heat he enabled us to take current electricity, electro-magnetism, etc., into the same category. In fact, even in 1840, before

[1] *Phil. Mag.* 1843, II.

he had come to definite conclusions as to the generality of the principle of conservation, he had established experimentally a grand series of particular cases of it; and one of the most remarkable was this:—

When any voltaic arrangement, whether simple or compound, passes a current of electricity through any substance, whether an electrolyte or not, the total voltaic heat which is generated in any time, is proportional to the number of atoms which are electrolysed in each cell of the circuit, multiplied by the virtual intensity of the battery.[1]

Therefore, even at that early time, his experiments (and his reasoning was entirely based upon experiment) had led him to this conclusion, that whenever something that was imponderable disappeared, and there appeared some other imponderable which could have no other origin, then the quantity of the one was directly proportional to the quantity of the other, and the ratio between these two had only to be determined by accurate measurement in order that you might know the mechanical equivalent of so much current electricity, or of so much heat, or even of the potential energy of so much zinc and dilute sulphuric acid, or of any other substances in a state fit for chemical combination.

Another most valuable experimental research of Joule's bears on the question of the mechanical value of *Light*.[2] He compared the heat evolved in the wire conducting a galvanic current, when the wire was ignited by the passage of the current, with that evolved when (with an equal current, suppose) it was kept cool by im-

[1] *Phil. Mag.* 1841, II. p. 275. Paper read before the Royal Society, December 17, 1840.
[2] *Phil. Mag.* 1843, I. p. 207.

mersion in water. These experiments showed a small, but unmistakeable, diminution of the heat when light also was given out. However, all that was necessary in order to extend the principle of conservation to light was to show that light, like heat, electric currents, and so on, is a form of energy and not a form of matter; in other words, to establish what is called the undulatory theory instead of the corpuscular theory.

I may digress for a little to say a word or two as to how that was done. It is one of the important advances made within the period to which my lectures chiefly refer.[1] It was established in France by Fizeau and Foucault, working originally by independent processes, but afterwards working together. The proposition then to be decided upon is: Does light, as it comes to us from the sun, for instance, consist in the transference of particles of something luminiferous? Is it matter, in fact, which is shot out from the sun? or is it a propagation of disturbance of some kind or other which may be assimilated, for purposes of illustration, to wave-motion? Is it, in short, a propagation of energy in some form or other, whether wave-motion or not, or is it a propagation of matter? Now, Newton and Huyghens had, long ago, each from his own point of view, assigned

---

[1] I am aware that many excellent authorities attribute the establishment of the undulatory theory to Young and Fresnel—saying that interference as in the phenomena of diffraction, etc., had, in their hands, completely upset the corpuscular theory. But, as a fact, some of the more noted supporters of that theory (including Biot) were not convinced by these experiments, but were led to make further modifications of their favourite theory, while there can be little doubt that they would have accepted Fizeau and Foucault's results as decisive against them. Of course, such a statement as this in no way impugns the value of the magnificent work done by Young and by Fresnel.

the means of perfectly settling this question. Newton, in fact, had shown that if light be matter, then, on being refracted into a dense body, it will move more nearly in a direction perpendicular to the surface, provided it move faster in the dense body than in the rare one outside. That is to say, that, since we know that an oblique ray of light falling upon the surface of water, for instance, which is denser than the air, is refracted more nearly to the vertical, Newton had mathematically demonstrated that if light consist of particles, it must move faster in water than in air. Huyghens, on the other hand, showed that if light consist of wave-motion, and be refracted towards the vertical, at the horizontal surface of a dense body such as water, then its velocity in the dense body must be less than its velocity in the rare body. Thus there was a distinction of the most marked character between the two theories. If therefore you can discover by experiment whether the velocity of light is greater or less in water than in air, you settle for ever the question whether light consists in the propagation of matter or in the propagation of motion or energy. Now the experiments separately made by Fizeau and Foucault both gave the result, that *in water light moves slower than in air*, and therefore it necessarily followed that light is a form of energy.

So far, then, we have come to the complete establishment experimentally of the classification of the imponderables under the head of energy, and we have arrived at a general notion of relations of equivalence between them. The mere fact of conservation, of course, at once establishes that there must be relations of equivalence. So much of the one is equivalent to so much of the other, provided you can effect the conversion of the one into the other. Of course it will always, or at least for a very long time, remain an extremely difficult problem to measure the equivalent of an amount of light. Still, it has been approximated to, and, among other processes, in this way : Light, when absorbed by an opaque body, is found to make the opaque body hotter. Here is an example of the principle of conservation. The energy of the light is not destroyed, but its vibratory motion cannot pass through this opaque body as light. It is employed in agitating the particles of the opaque body, and that body becomes hotter in consequence. We can measure, then, the quantity of light in terms of the heat which it produces, or to which it is equivalent, and then we can measure that quantity of heat in terms of mechanical work, so that, as Sir William Thomson did many years ago, shortly after Joule's discoveries appeared in print, we can calculate what he calls the mechanical value of a cubic mile of sunlight; we can calculate how many foot-pounds of work are equivalent to the sunlight which a cubic mile of the earth's atmosphere, filled with direct sunlight, has in consequence of that luminous energy which is passing through it at the instant.

Before I leave for the moment the subject of the conservation of energy, I must speak of one additional name in connection with its discovery and early development, that of Helmholtz, the great physiologist of Berlin, who has now, at least nominally, ceased to be a physiologist, but who remains one of the foremost of living mathematicians and natural philosophers. One of his early works was published in 1847, shortly after Joule and Colding had published their discoveries. It seems, however, that he was barely acquainted with the writings of either, but had set to work himself, from a mathematical point of view, to settle the principle of conservation of energy. In fact, the German title of his book is precisely an equivalent to our English phrase 'conservation of energy.' He based the principle upon one or other of two propositions, and it is interesting in the highest degree to consider what these propositions are, and to see how a man who was fully acquainted with the whole science of the time looked at a subject of this sort, and pointed out in what direction experiment ought to be turned in order to verify the conclusions of theory. He says, in effect, that if you take Newton's principle—the principle you have already heard [p. 33]—and if you combine it with one or other of the two following postulates, you will establish completely the conservation of energy. The first postulate is: Let us suppose matter to consist of ultimate particles which exert on each other forces whose directions are those of the lines joining each pair of particles, and whose amounts depend simply on the distances between the particles. Suppose, in fact, that something akin to gravitation-force exists amongst all the particles of matter in the universe, that each particle attracts every other particle with a force which depends only upon the distance between them, not in any way upon the sides which are turned to one another, so that if you know the distance between them you know the amount of the attraction, and that the attraction shall also be (in accordance with Newton's Third Law of Motion) in the direction of the line joining them. If you make that assumption, then it is a mere consequence of the ordinary laws of motion of gross matter that, if all forms of energy depend upon motion or position of such particles, the conservation of energy must hold, and also that the so-called perpetual motion would be impossible under any circumstances.

As an alternative, Helmholtz shows that we may take as our postulate this consequence of the first postulate. Take the impossibility of the so-called perpetual motion as a postulate, and take along with it Newton's grand statement of his second interpretation of the Third Law of Motion, these two together would, by themselves, enable you to prove the principle of conservation of energy. Now it had for many years back been an accepted matter among men of science (as typified by the long-since announced determination of the French Academy to consider *as not having reached it*, any paper whatever upon the perpetual motion), it had been accepted by men of science, I say, almost universally that experiment had conclusively demonstrated the perpetual motion to be impossible. So Helmholtz, by showing that if you simply begin with that experimental fact, and take in addition to it Newton's statement, you can establish the conservation of energy, had made, independently of Joule and Colding, a discovery of this great principle for himself. You will notice that he did, almost as distinctly as either Joule or Colding, insist upon the necessity of experiment for the establishment of such a principle, but he brought in his experiment in the form of an universally accepted result of the experiments of others, namely, the impossibility of the perpetual motion, while they preferred to make perhaps more direct experiments for themselves.

I shall have occasion to say a word or two more about the so-called perpetual motion, because it has really been for natural philosophy—and it remains even to this day—as important in its influences, especially in aiding us to simple proofs of important theorems, as, for instance, the notion of alchemy has been in chemistry. We all know that if there had not been a pursuit after the philosopher's stone, chemistry could not yet have been anything like the gigantic science it now is. In the same way, we can say that modern physics could not yet have covered the ground it now occupies had it not been for this experimental seeking for the so-called perpetual motion, and the consequent establishment of a definite and scientifically useful negative.

We notice, then, as a deduction from what I have just explained about the work of these three independent discoverers of conservation of energy, that all physical phenomena are necessarily transformations of energy of some kind or other ; and we may carry our deduction so far as to say that even that mysterious thing, whatever it may be, the life of plants and animals, is, so far as it is physical, entirely an exhibition of transformations of energy. There are things connected even with life which may not be purely physical. There are other things associated with living beings which, of course, no one in his senses can regard as physical. Even such things as *Consciousness* and *Volition* we have absolutely no reason, however vague, for classifying, even in the smallest degree, under the head of physics. But everything which is really physical in life—and we are beginning to find many things that are so—is merely an example of some form of transformation of energy.

Having said so much, it will be obvious to you that our proper course now will be to consider the principle of transformations, and then inquire in what direction we must seek for more light. We shall find that the question which is suggested by all these tentative experiments is, What is the law of transformation of energy? From a given quantity of a given kind of energy, how much of another assigned kind of energy can be produced by a given process?

This question breaks up into two. The first is, How much of a given kind of energy can be transformed into some other given kind? And then there is a second question: When you have got so much of it transformed, to how much of the other kind will it correspond? That is the question of equivalence again. I have already discussed that, so I confine myself now to the first question. The first question fully stated is: Given a certain quantity of energy in one form and under given conditions, how much of it can you, by means of a given kind of apparatus, convert into some other definitely assigned form, the rest being either untransformed, or transformed in whole or in part into some third form? Now, you will see at a glance that there is something very important under this. Just think for a moment of the enormous amount of waste which is known to take place in an ordinary steam-engine. In the very best engine, even if it were theoretically perfect, and working at ordinary ranges of temperature, it has been satisfactorily demonstrated that only somewhere about one-fourth—very rarely so much as that, but at the best about one-fourth—of the heat which is actually employed is converted into work; that is to say, three-fourths of the coals, or three-fourths of the heat employed, are absolutely wasted under the most favourable circumstances. Now, what is it that determines this? Why is it that if I have a quantity of work or potential energy I can convert the whole of it, if I please, into heat; but when I have got it converted into

THE CONSERVATION OF ENERGY. 73

heat, I cannot convert the heat back again, except in part, into the higher form of work or potential energy? The answer is included entirely in that word '*higher*,' which I have just used. When you are converting energy from the high form into the low, you can carry out the process in its entirety, but when it comes to be a question of the reversal—going up-hill as it were—then it is only a fraction, in general (even under the most favourable circumstances) only a small fraction, of the lower kind of energy which can be raised up again into the higher form. All the rest sinks down still lower in the process. When you have got it low already, and when you are to elevate part of it and transform it into a higher order, you must inevitably still further degrade a large part of it; in general the larger part of it. This, as we shall find later, is one of the most important scientific discoveries ever made: having most stupendous bearing on the future of the whole visible Universe.

I shall conclude this lecture by showing some examples of conservation of energy with the apparatus before me. I shall necessarily at the same time give some illustrations of transformation of energy, independent altogether of the particular physical experiments which are employed for the purpose. I am merely giving you these experiments as illustrating conservation of energy, and incidentally, in addition, transformation and dissipation of energy, so that we are not concerning ourselves with what is the branch of physical science to which any particular experiments belong, but simply with how far the experimental results help to illustrate the transformation.

Take, then, first of all, the simplest form—the case of

an ordinary pendulum. When the pendulum is vibrating, there is constantly going on transformation of energy of the very simplest kind—transformation from the potential form which I give it by drawing it aside (and therefore lifting it), and which it gradually loses as it falls back, getting more and more kinetic energy instead, until at the middle of its course, when it is moving fastest, it has its greatest amount of kinetic energy, having lost for an instant all its potential energy. Then it gradually loses the kinetic energy as it is climbing up again, and regaining potential energy, then the energy is all potential, then it becomes kinetic again, and so on. Of course if there were no air-resistance, and if the stand itself were absolutely rigid, and the cord supporting the mass flexible and inextensible, this process would go on absolutely for ever. It would be perpetual motion, but it would not be *the* perpetual motion. Remember the distinction there. Perpetual motion is simply a statement of Newton's First Law of Motion. All motion is perpetual until force interferes to alter or modify it. But this is not *the* perpetual motion, because, although under the favourable circumstances I spoke of just now, the pendulum would remain for ever moving with the same quantity of energy it has at present, yet it could not help you to drive machinery, except at the expense of that energy. It cannot drive anything else without losing part of its own energy, and when that occurs, the case does not come under the head of what is called *the* perpetual motion, although, when there is no drain upon it, it may be *a* perpetual motion.

Now, as we know by experience that this vibration will not go on for ever, let us consider why it is that its energy is gradually being lost. What becomes—and, according to the principle of conservation of energy, we ought to be able to trace it—what becomes of all the energy I gave it at first? Well, we see in a short time that it is communicating motion to the air around it; every time that it vibrates backwards and forwards it sends alternately a wave of compression and one of dilatation through the air of the room. These waves do not sufficiently rapidly succeed one another to produce an impression upon our sense of hearing, but they are sufficient to agitate the air of the room. They are propagated through the air of the room with the velocity of sound, and they are gradually frittered down into heat because air is not a perfect fluid. Because then there is something producing effects akin to those of friction amongst its particles, these waves are gradually rubbed down into heat, and if we had a sufficient number of such pendulums set into vibration to begin with, and all sufficiently resisted by the air, we should be able to warm the air of the room, no doubt to an extremely small extent, but still so that the quantity of heat produced should be precisely equivalent to the quantity of energy which you had communicated to the pendulums at starting. But then this suggests another question. At present the pendulum, hanging at rest, has no potential energy, that is, if the string cannot be cut. It has at present potential energy if you can cut the string, because it will drop on the table, or at least it will have the power of falling. But suppose the string is absolutely inextensible, and cannot be cut, then we must consider it in this position as having no potential energy at all, because it cannot get down any lower than it is at present. How is it, then, that I can give it energy? because if there be conservation of energy, and if we so put it that it has none to begin with, and it gets some, there must be some other energy spent in communicating it. Now, that leads us to the grand consideration of the source of animal energy, because, by pressing the pendulum with my hand, and thus elevating it, I must have done work, for I have exerted a pressure through a certain space. Work has been done, and therefore something has been expended in my body for the purpose of producing it. This raises the question of how the animal supplies the work; and the further one, in what form does the animal get the work supplied to it, which it is constantly giving out even when in repose? Of course you can at once see that it must be in some way or other connected with food. That, then, will lead us, in another lecture, back to the consideration of whence the food derives its energy, and so on in succession. So you see that even so simple an experiment as setting this pendulum in vibration leads us to a train of consequences, both back and forward, in reasoning, which might well occupy us for a whole series of lectures. Nothing is better calculated to show at once the profundity of Nature's secrets, and the firm grasp we have already taken of some of them, than an example like this—so simple and yet so complex.

Instead of taking the case where the motion of the air is not capable of being perceived by the ear, let us take a case in which we use a special instrument for the purpose of communicating vibrations to the air in such a form that the ear can seize them. If I were to take this tuning-fork and strike it against the table, or start it in any of the ordinary ways, and it were not provided with this sounding-board, the amount of surface which it presents to the air is so slight that the amount of energy which it would spend in a given time in the form of sound would be exceedingly small; and therefore the sound would be hardly audible at any considerable distance. But when we furnish it with a resonant cavity, as it is called, such as this, every part of which is set in vibration by the motion of the fork in exactly the same period as the fork; and when, moreover, the dimensions of this cavity containing air are exactly adjusted, so that when it is set in vibration, it tends to vibrate in exactly the same time as the fork, then we have got a sensitive apparatus which enables us, as it were, to lay hold of the air, and to dissipate or spend at a very great rate the energy which we give to the fork. The pendulum here spends it at a very slow rate, but in this fork we have applied our knowledge of physics so to construct an apparatus as to make it spend its energy or communicate it to the air as rapidly as possible. We have it now in the form of sound affecting our ears, but you will notice that the sound gradually dies away. The vibrations of the tuning-fork die away far faster than those of the pendulum, because if you will give out the energy at a great rate, the original stock can last only for a short time. The greater the rate at which you give it out, the shorter the time for which it will last. But there is another cause in this case for the very speedy cessation of the sound. The greater part of the energy which I gave to the tuning-fork by muscular work done in forcing these prongs asunder for a moment, the greater part of that energy is spent in heating the body of the fork itself. Steel—however startling this may appear to some of you—is exceedingly imperfect in its elasticity. When a steel bar, such as this, is rapidly changing its form, there is an enormous amount of internal friction, and thus is consumed a great part of the energy which is given to it, so that only a part of the energy originally communicated is given back in the form of sound, even with the help of the resonant cavity.

To take another instance. I have got a galvanic battery under the table, and it is connected with a certain electrical apparatus. Now, whenever I allow the electric current to pass through this apparatus, there is for the moment a certain quantity of zinc consumed, or, as we may put it, a certain quantity of potential energy in the battery has been converted into the kinetic energy of a current of electricity. That current of electricity passes round some yards of copper wire, coiled round a bar of iron or a number of fine iron wires which are standing vertically inside this apparatus. The moment the current passes, these iron wires are converted into magnets, but, in consequence of the conservation of energy, while this is going on they weaken the current. The current of electricity becomes weaker in the act of making the magnet, but the moment the magnet springs into existence it again is weakened, because, from the necessities of its position, its mere coming into existence necessitates the passage of a new current of electricity in another coil of wire which surrounds this externally. So that here are a number of transformations: First, we have a certain amount of zinc dissolved, *i.e.* a certain amount of potential energy lost; then a certain current of electricity produced in consequence; then that current of electricity weakened by producing magnetism in certain iron wires; then the magnetism of these iron wires re-acted upon to produce a new current in another set of wires; and finally, we can use that induced current, as it is called, to produce heat, or light, or sound. Let us try it, for instance, in such a form as to produce heat. Every time you hear that click [of the contact-breaker], a fresh amount of zinc has been dissolved, and in consequence that series of transformations I have just described has taken place. You will notice that the zinc is burning, though without almost any development of heat, in the battery, but we can have the fire wherever we please. We have no heat, at least nothing to speak of, in the battery. The heat that would be produced by the dissolving of the zinc is not developed inside the battery at all;—if we had a couple of Atlantic cables here, between the battery and this apparatus, we should be able to produce it at a distance of 3000 miles from the place where the fire burned. In order to show that heat is produced largely in such a case as this, my assistant will hold a piece of paper between the poles. [You see it is at once ignited.] You will notice that the burning of the zinc is below the table, but it might have taken place 3000 miles off if we had had good enough conductors. There you see it has at once produced a development of heat sufficient to inflame the paper. Now, I may easily alter this in a striking manner. Use the same amount of zinc as before, or as nearly as possible the same amount of zinc, but instead of the spark being a quiet one, make it noisy and luminous, as you see is easily done by attaching the coatings of a Leyden jar to the ends of the secondary coil. Then we shall find that it is not so hot as before (at least so far as the paper test can inform us). Of course it could not be expected to be so hot, because, if conservation of energy be there, and if there is a certain quantity only of energy that the spark can have, and if it be made to spend the greater part of that energy as sound and light, you cannot expect it to have as much heat as before. You see it now immensely brighter than before, and accompanied by a sharp crack, but we might go on with the experiment indefinitely, and never set the paper on fire.

This is a very excellent instance of multifold transformations, and furnishes also, as you have seen, a rough illustration of conservation.

LECTURE IV.

TRANSFORMATION OF ENERGY.

Experimental Illustrations—Heating of wires, and decomposition of water, by a Galvanic current—Electro-magnetic Engine—Rotating Disc—Magneto-electric Machine—Induction-Coil and Geissler Tube—Higher and Lower Forms of Energy. Work transformed wholly into Heat—Only a portion of the Heat can be reconverted into Work. Carnot's Cycle of Operations and his Reversible Cycle. Effect of pressure upon Ice.

IN my last lecture I showed you how, mainly by Joule's grand experiments, it had been conclusively demonstrated that conservation holds for every form of energy, and therefore that all physical phenomena consist in mere transformations of energy. There cannot be a destruction or creation of energy. All that we can have is a modification or transformation of it; and therefore we must to-day consider more fully the laws of such transformation. I shall begin the consideration of them by taking one or two experiments, and pointing out in each of them the various forms in which the energy appears,—how it was first introduced into the apparatus, under what successive forms it passed through the various parts of the apparatus, and in what final forms it was thrown out.

Galvanic Battery with stout copper terminals.—The first and simplest experiment of this kind is the production of heat directly by chemical combination. As in all or most of the experiments I am about to show,

I intend to begin with a galvanic battery, I may say a word or two as to the form in which its energy appears. The energy in the battery consists mainly in the fact that we have zinc which is capable of being burned, as it were, by being dissolved in dilute sulphuric acid. Now, if we were to burn the zinc, as can easily be done by simply allowing it to dissolve (that is, by not taking the precautions we have here taken against its dissolving without permission in the sulphuric acid), we should, simply in consequence of the potential energy which is lost by the zinc and the acid when they combine, have a certain amount of heat generated by their combination, and this would be developed in the cell of the battery. But instead of permitting this, we can cause the combination to take place without almost any development of heat. We can have practically all of it in the form of some other manifestation of energy. We can have it in the form, for instance, of current electricity; and we can employ the kinetic energy of that current for the purpose of producing various other forms of energy by suitable transformations. In consequence of the amalgamation of the zinc, and the other precautions taken in the cells of the battery, very little combination goes on in this battery until the circuit is *closed*, as it is called; but as soon as we close the circuit, by joining together the terminal wires, a current of electricity passes. A current of electricity is now passing through the circuit, and chemical action (both decomposition and combination) is going on to exactly the same extent in every one of the cells. But the chemical action now going on is attended with the development of a large quantity of heat in the cells, almost precisely the same amount of heat as would have been developed if we had dissolved the same quantity of zinc in the sulphuric acid without any production of electricity at all; the reason being that the conducting power of this wire which I have for the moment used to close or complete the circuit is so great that the small resistance it offers to the electricity scarcely fritters any of the electricity down into heat. The heat which is equivalent to what would be produced by the direct burning of the zinc, is all or almost all produced in the cells themselves, because it is in *them* that the current suffers resistance. But if I interpose in the path of the electricity an imperfect conductor, which shall resist a great deal more than the copper wire, or even than the cells themselves (as I do by inserting in the circuit a long fine iron wire), then you notice that we get the heat (which is really due to the chemical action taking place in the cells),—we get that heat produced in another locality altogether, and we could have transferred that locality as far away as we pleased, if we had simply made our copper wires thick enough and long enough. By simply making them thick enough, so as to waste as little as possible of the kinetic energy of the current electricity, by friction on the way, we should have kept it all or nearly all for the purpose of developing as far as we please from the battery the heat really due to the combustion there.

Voltameter introduced in circuit.—Instead of using the current electricity for the purpose of producing heat, let us endeavour to ascend again from the kinetic energy of the current to potential energy of combustibles. Remember that it was the chemical potential energy of combustibles which we had in the battery to begin with. By allowing the zinc to dissolve, we got our current electricity, and now we shall use that current for pulling asunder two substances in chemical combination. We shall use it simply for the purpose of decomposing water. By causing the current to pass through a vessel of water, you notice that we cause bubbles of gas in large quantities to ascend from the ends of the conducting wires ; and we have the kinetic energy of the current spent entirely, or almost entirely, in pulling asunder, against their chemical attraction, the particles of oxygen and hydrogen which form the water. You see that a quantity of the water is being decomposed, for you see how the gas is bubbling up through the water from the end of this collecting tube. Now, supposing there to have been no loss during the operation—no frittering down of the electricity into heat—but that the whole energy of the electric current has been spent in decomposing the water, then the potential energy of the separated oxygen and hydrogen which I collect in this way should be precisely equivalent to the amount of potential energy which was consumed in the battery, or rather was there transformed into the energy of the current. In order to show (with as little risk as possible) that there is a large amount of potential energy in these mixed gases, all we have to do is to employ them to produce froth in the form of a multitude of small soap-bubbles blown with the mixture. By applying a lighted match, we shall be able to produce from the potential energy of the mixed gases a violent explosion, which of course represents a certain amount of energy. That explosion gives you light, heat, and a very loud sound. The sum of all these energies taken together, provided nothing has been lost during the process—that nothing has been frittered away (by breakage of the mortar, for instance)—will represent precisely the amount of energy corresponding to the amount of zinc which has been dissolved during the operation. You notice that here we have now in another form—and a form which affects the air more than any of the other forms of energy we have used—the energy which ought to have been developed in the form of heat by the combustion of the zinc, but was not, because we had electricity in the place of it; then, in place of that electricity, we had work done in overcoming the chemical attraction of oxygen for hydrogen; then we had the mixed gases, which as soon as we pulled the trigger, as it were, by applying the lighted match, gave us back our energy in another kinetic form, or as a mixture of several kinetic forms.

Electro-magnetic Engine.—You had in the voltameter current electricity produced by the battery, and employed for the purpose of producing potential energy, by separating the particles of a chemical compound. But we can produce potential energy by the help of a battery by another and somewhat simpler method. Suppose we employ the current of electricity produced by the same battery, for the purpose of setting an electro-magnetic engine at work. (We are not at present concerned with the details of construction of the engine.) For this purpose we do not (at least with the engine before you) require anything like so powerful a battery as we used for the rapid decomposition of water. Two, or at most three, cells will be sufficient for our present purpose. You notice that the current is now producing motion of machinery, and has actually raised a weight—not by any means a great one, but still the fact remains that a certain mass has been raised against the earth's attraction to a certain height above its surface; and you can easily see that, if the experiment succeeds through a space of three or four feet, as it has now done, it would equally succeed (if we kept the engine working long enough) in enabling us to raise the weight, by proper mechanical adjustments, to any height whatever. Now, let us consider what transformation of energy took place as the current of electricity passed round these electro-magnets, being shunted now into one of them and then off it and into the next; into each when its becoming a magnet will aid the desired effect; off it when it would tend to hinder it. This is a mere detail of mechanical arrangement, and is effected by different combinations of machinery in different electro-magnetic engines. But we are not concerned with details of machinery; we confine ourselves to the transformations of energy which are going on during the working of the engine. But from this point of view what takes place here? The energy of the current is to a certain extent converted into the raising of weights; that is to say, potential energy is produced in place of the kinetic energy which was supplied from the battery; but if the current not only drives this machinery but keeps it doing work, then there would not be conservation of energy unless the current itself were kept at a reduced strength, at least while it is in the act of doing work. Now, that is what is found to take place. It is found that while the engine is working, the current is considerably feebler than it is if we were simply to stop the engine, and allow the current to pass without doing any work. This is quite analogous to the case I pointed out to you in a former lecture. When a given quantity of steam is blown through the engine from the boiler into the condenser without doing any work, we find that the quantity of heat which goes into the condenser is larger than the quantity of heat which goes into it while the engine is doing work. In precisely the same way, then, while the current of electricity is employed in actually lifting a weight, or in driving an electro-magnetic engine, the current which is passing along the wire is feebler than before, and corresponds, according to a great discovery of Faraday's, to a less amount of chemical combination (that is, a less rapid consumption of zinc) in the battery. The battery has really less hard work, while driving this electro-magnetic engine, than it would have if we were simply to stop the engine and allow the current to pass and develop heat in the conducting wires and cells. It must do something. The current of electricity always fritters itself down into heat in time, unless you utilise it and change it into a form of energy more useful than heat. But what we find is this, that, though there must of course always be a current passing:—or else these iron horse-shoes would not successively become electro-magnets—the current is very much weaker when the engine is doing work than when it is not. And it is also found that the weaker the current becomes (the more the current is checked by reflex action, as it were,—by the resistance it meets with in doing work), the greater is the percentage of the amount of energy really spent in the battery which is finally converted into useful work. Thus, in order to get an electro-magnetic engine of this kind to do work on a large scale and at a profitable rate, it would be necessary to drive it with enormous rapidity; for the faster it is driven the greater is the reaction upon the current, and therefore the more is the current enfeebled, and the greater the percentage of the driving power which is utilised. And the laws discovered by Faraday and Joule respectively—viz., that *the strength of the current is directly as the quantity of zinc dissolved per second*, and that *the heat developed is directly as the square of the strength of the current,*—show that the efficiency of the engine is directly proportional to the weakening of the current. The more the engine weakens the current by reaction, the greater is the fraction of the whole amount of fuel spent which is converted into useful work.

Many of you are doubtless practically much better acquainted with the subject I am now to mention than I am, and therefore I shall only briefly state that, even if we could succeed in making an engine of this kind work at a very great speed, and thereby obtain the highest efficiency possible ; and if we could, for the purpose of keeping up such a speed, almost wholly get over the difficulties of ordinary friction, which of course become far greater and more serious as the rapidity of the working of the engine increases,—even if all this could be done, still, if we calculate the cost of the fuel here, we shall find that such an engine could never economically compete with an ordinary steam-engine, because of the fact that in order to smelt a quantity of zinc, an expenditure of about sixty times its weight in coal is required ; while, weight for weight, the coal is far the more powerful fuel, *i.e.* loses far more potential energy in being burned ; and therefore of course there can be no comparison between the prices of the fuel in the two cases, if the same ultimate amount of work is done.

Copper disc with multiplying gear.—The next case I take is a very curious one. I have got here an arrangement (never mind the details) consisting of a driving wheel and multiplying gear, by which I can communicate an extremely great velocity of rotation to this copper disc, which is mounted as freely as possible upon well-oiled and well-supported axles. It is, in fact, easily driven at a rate of somewhere about a couple of hundred turns per second, if we work the driving handle at the rate of about two turns per second. The disc consists of a highly conducting material—copper, and it is placed between two pieces of iron which do not touch it, but come very near it. These pieces of iron form part of the armature of a small electro-magnet. Now, the coils of this electro-magnet have at present no current passing through them, and I find that, as you see, there is nothing more easy than to set the disc in very rapid motion indeed. You notice that when I remove my hand, the inertia of the wheel-work is such that the whole goes on turning for a very considerable time. Now notice what the effect will be if, while I am driving it, my assistant suddenly throws the current, even from three cells of a battery, round the electro-magnet. Then I shall be endeavouring to drive the copper disc in the immediate neighbourhood of a strong north pole on the one side of it, and an equally strong south pole on the other. Although there is no contact —nothing of what we ordinarily call friction—you will see that this acts exactly like a friction brake of very great power. There; you observe the instantaneous stoppage, and you also see that, strive as I may, I can scarcely move the driving handle. With such battery power as that, it is utterly impossible for any one man to drive the disc fast; it would require perhaps four or five persons to force it to rotate at even a very moderate speed. If I put on a single cell instead of three, you see that by great exertions I manage to keep the disc rotating at a slow rate for a short time; but it is only by the expenditure of a very considerable amount of labour. I could keep it going perhaps for a few minutes, but there is no necessity for pushing the trial further. Now comes the question, What have we to show for this? What necessitates the extraordinary amount of effort that is required in order to keep the disc turning in the magnetic field? In order that you may see this experiment in another and perhaps a clearer light, I shall take advantage of the fact that, as you saw a little ago, the machinery is capable by its inertia, if once set rapidly in motion, of going on for a considerable time before the motion finally dies out. I start it again, with the same rapidity as before, and you see the almost instantaneous collapse as soon as the circuit is closed. We have in fact a friction brake acting without contact, and to force that disc to move rapidly in the neighbourhood of the magnet requires an enormous expenditure of work. Now comes the question, Where does this work go to? Suppose that in spite of this enormous resistance to the motion of the disc, we were to expend work in turning it. The answer must simply be this, that the whole, or almost the whole of the work so spent goes to heat the disc: and that, simply by persistently turning it under these circumstances, you can make the copper absolutely red-hot, and, in fact, melt it, if the experiment is carried on far enough, without any contact whatever with the iron of the electro-magnet. The mode in which this heat is produced is also very interesting. It depends upon induced currents, one of Faraday's great discoveries. Faraday discovered, as I daresay you are all aware, so long ago as 1831, that when a conducting body is made to move in the neighbourhood of a magnet, the relative motion of the two produces currents of electricity in the conductor. Now, when a current of electricity is once produced, we have seen that unless it be diverted to produce work, or potential energy, or some other form of energy, it always in time fritters itself down into heat. If, then, you keep this copper disc moving in the neighbourhood of the magnet, the faster it moves the stronger are the currents produced in it; and as there is no appliance here to collect these currents, so as to utilise them for any other purpose, the currents must fritter themselves away into heat in the copper disc itself. A permanent magnet would have precisely the same effect as our electro-magnet—the only reason for using an electro-magnet being that it is so easy to magnetise and demagnetise the soft iron, *i.e.* virtually to present or withdraw the magnet by the mere making or breaking of contact of two wires. The currents which are generated in the disc, are in such a direction as always to be attracted by the magnet; or, as it may be more scientifically put, in the words of Lenz, the mutual action between the magnet, and the currents generated by the relative motion of the conductor, always tends to diminish that relative motion. Hence the work constantly required to maintain the rotation of the disc.

Magneto-electric Machine.—Now, still further to illustrate this part of the subject, I may refer to this magneto-electric engine, which was devised to take advantage of Faraday's discovery just mentioned. Here are a couple of coils of wire with iron cores, which are to be made to move in presence of a bundle of steel magnets. Here we have, in a somewhat different shape, the essential features of the engine I have just been using. We apply a certain amount of mechanical work, in order to move these coils in the presence of the poles of the magnets; and thus have currents developed in them as we had them developed a little ago in the simple copper disc. I am now about to collect these currents for the purpose of producing light, instead of allowing them to be frittered down into heat, as in the former apparatus; and you see that we produce a brilliant spark by simply expending mechanical power or work upon the driving handle, without any battery, without any electro-magnet, or anything of that kind. By simply forcing the conductor to move in presence of the steel magnets, we can develop currents strong enough to produce that brilliant spark. Of course with this little machine the light is on a very small scale, but the engine is acting on precisely the same principle as the magneto-electric machines, driven by steam-power, which have been recently employed with great effect for the purpose of lighthouse illumination.

Induction Coil with Geissler-tube containing highly rarefied Carbonic Acid.—There is only one other illustrative experiment connected with these to which I shall now advert, and that is another mode of converting work or potential energy into light; that is, by means of an induction coil, as it is called. I am using with it the battery I have hitherto been employing. We produce a current of electricity by means of it; we magnetise a bundle of iron wires by the help of that current; then we break the circuit and stop the current, and the iron wires cease to be magnets. At the instant that they cease to be magnetic they are virtually, as it were, suddenly pulled away to an infinite distance. Now, this coil (consisting of a very long conducting wire) is in the immediate neighbourhood of the bundle of iron wires. When they become magnetic, it is as if a powerful magnet were suddenly inserted in the coil. When they cease to be magnetic, it is as if the magnet were instantaneously withdrawn. In either of these cases, we have the development of an electric current in the conducting coil. Now, instead of driving that current through a very small space of common air, as I did in the case of the magneto-electric machine, I will drive it through a considerable length of the contents of a highly exhausted receiver. I do this for a particular reason, which will appear as soon as we have got the room darkened. You now notice the exquisite luminous effect produced by resistance: but observe especially this peculiarity about it, that it remains persistent for a certain time after the discharge has been interrupted. You see at once that the discharge has ceased, by the disappearance of the purple and the blue light near the ends of the tube; while the olive green light which is in the wider parts of the apparatus remains for a time visible, and gradually dies away. It has scarcely yet, as it were, cooled. It presents, except as to colour, exactly the appearance of a heated body cooling. This remarkable effect then, though due primarily of course to the current, gives us a curious instance of a body which, when agitated by the passage of the current, can convert its energy into light, and part with it in that form. There is in fact scarcely any radiation of dark heat from that glowing and cooling body. I interpolated that experiment just now, not because it has any direct connection (except as to the exciting cause, the battery) with what we have had before, and shall have immediately after it; but because I had the apparatus ready, and it was as well to show the experiment while it was at hand.

In all these cases you will have noticed that there has been a transformation—sometimes many transformations in succession; but there is one law of nature which we notice in the case of all these transformations. Some kinds of energy are of a higher order than others, and if you begin with one of the higher orders, you can get from it any of the others, and in general you can transform almost the whole of it into any of the others you please; but when you begin with one of the lower forms, the reversal of the process is attended by extraordinary difficulties. The lines

> . . . facilis descensus Averno;
> noctes atque dies patet atri janua Ditis:
> sed revocare gradum, superasque evadere ad auras,
> hoc opus, hic labor . . .

seem almost to have been written by one who anticipated our knowledge of the laws of the transformation of energy.

We come then to the question of the raising of energy from lower to higher forms, which is the only one which presents much difficulty; and if we thoroughly understand upon what conditions the utmost transformation of heat into work depends, and how it is that at best only a small fraction of a given quantity of heat can, under the most favourable circumstances, be converted into work, then we shall have no difficulty whatever in seeing that laws of a similar kind, although not perhaps precisely the same, must hold for every other transformation from one form of energy to a second, especially if the second be the higher form of the two. Now, the ordinary conversion of work into heat you may see illustrated in the most direct form in manifold ways. Savages, for instance, procure a light by rubbing two pieces of dry wood together, or still better by using a piece of hard wood to bore a hole in a soft piece. Any of us can effect that operation, and set the pieces of wood on fire, by applying long enough and with sufficient rapidity and pressure a sort of drilling motion. It is quite easy, by the expenditure of a little mechanical energy, to set fire to both pieces of wood. That is merely of course an improvement upon the apparatus used by the savage. When we stir or churn, or anyhow rapidly agitate a mass of water, we find that the amount of work we spend upon it is at first converted into actual or kinetic energy of the moving water. You see it rotating round as you stir the vessel; but if you leave it to itself, you see that its rotation gradually slackens until it comes finally to rest. In such a case, it is found that the whole of the work spent upon the water has been ultimately converted into heat. Whenever you apply work to the production of heat by friction, you have an apparatus perfect enough to get the whole of the work transformed into heat. It may be that part of the energy is originally not in the form of motion, as when part of the surface of rotating water is raised above its mean level, but this potential energy also gets frittered down into heat by degrees. It may be also that, even in ordinary friction, even in such a case as the friction of sand-paper against a piece of wood, the first thing produced by the friction, or rather by the work spent in friction, consists of electric currents in the immediate neighbourhood of the place where the rubbing is effected. We have something very similar to that, although on a more delicate scale, in the case of an ordinary friction electrical machine. There is no doubt that the electricity there is produced by something very closely resembling ordinary friction, although it may be something intermediate between it and contact; but this leads us to the supposition that it may be possible that in many cases of what appears to us to be downright friction, perhaps even (as Sir W. Thomson says) when actually carried to the extent of abrasion of particles of the two bodies which are rubbed on one another, there may be, first of all, the production of electric currents to a certain extent, and that these currents may be almost immediately frittered down into heat by the resistance or bad conducting power of the two rubbing bodies; so that in such cases work spent in friction may not immediately produce heat. But there is no question whatever that whether heat be immediately produced or whether it is produced mediately, through electric currents, we can convert the whole of the amount of work spent in friction into heat.

Then in the same way we know that, by hammering a horse-shoe or other small piece of iron on an anvil, a skilful smith can without much trouble raise it to a dark red heat. The work spent in producing these impacts is almost entirely converted into heat, and this mainly in the piece of iron to which he applies his blows. And you will see something of the same kind, though on a grander scale, in artillery practice. Whenever the huge projectiles of the modern great guns have been employed for the purpose of penetrating armour plates, though a great part of their energy has no doubt been spent in actually penetrating the thick iron plate, yet at the same time there is an immense flash of light, accompanied by heat and various gases produced from the two metals by actual fusion and evaporation, all taking place at the instant of the impact, and corresponding to portions of the work transformed. In these cases, then, there is no difficulty whatever in getting the work converted directly into heat.

But we now come to the question how to get heat converted into work, and here our difficulties begin. Even in the best steam-engine, we cannot convert into useful forms more than between one-fourth and one-third of the heat which is employed.

In treating of this subject, I must introduce an advance in scientific method which was not known to men of science till within the last thirty years, although it was published in 1824; the great work of Sadi Carnot, a work of which it is impossible to speak in sufficiently high terms in such a series of lectures as I am giving. I need only say that without this work of Carnot's, the modern theory of energy, and especially that branch of it, which is at present by far the most important in practice, the dynamical theory of heat, could never have attained in so few years its now enormous development. Carnot's claims to recognition are of an exceedingly high order, because they depend not merely upon his method :—which is one of startling novelty and originality, and is not confined to the subject of heat alone :—but upon the fundamental principle on which he based his mode of comparing the heat employed with the work procured from it. Every reasoner (who has applied himself to the subject of heat since Carnot) has gone right, so far as he attended to Carnot's principle, but has inevitably gone wrong, when he forgot or did not attend to it. The fundamental blunders of Séguin and Mayer and various others—whose admitted claims I have pointed out in a former lecture—are almost entirely due to their ignoring the great principle laid down by Carnot so early as 1824.

Carnot's work is upon the *Motive Power of Heat*. It forms no inconsiderable portion of Sir W. Thomson's many scientific claims that he recognised at the right moment the full merits of this all but forgotten volume, and recalled the attention of scientific men to it in 1848; pointing out, among other things, that it enabled us to give, for the first time, an *absolute* definition of *Temperature*. Although Carnot (seemingly against his own convictions)[1] reasons on the assumption that heat is matter, and therefore indestructible; and although, in consequence, some of his investigations are not quite exact, his work is of inestimable value, because it has furnished us, not only with a correct basis on which to reason but, with a physical method of extraordinary novelty and power, which enables us at once to apply mathematical reasoning to all questions of this kind. These then are his two great claims,—first, the setting thermo-dynamics upon a proper physical and experimental basis; and, second, in the furnishing us with a means of reasoning upon it which was absolutely new in mathematical physics, and which has been, not merely in Carnot's hands, but in the hands of a great many of his successors, as fruitful in new discoveries as the idea of the conservation of energy itself.

---

[1] [*Note to Third Edition.* Since the publication of the last edition of this work Carnot's posthumous papers have been issued, along with a reprint of his great work. They indicate an amount of insight into the true theory, and the proper modes of experiment, truly marvellous even in comparison with the grand advances made in that work itself.]

Now, these two grand things which Carnot introduced, which were entirely originated by him, and which left him in an almost perfect form, were the idea of a *Cycle of Operations*, and the further idea of a *Reversible Cycle*.

In order to reason upon the working of a heat-engine (suppose it for simplicity a steam-engine), you must imagine a set of operations, such that at the end of the series you bring the steam or water back to the exact state in which you had it at starting. That is what Carnot calls a cycle of operations, and of it Carnot says, then, and only then, *i.e.* at the conclusion of the cycle, are you entitled to reason upon the relation between the work which you have acquired, and the heat which you have spent in acquiring it. If you were to take, as Séguin proposed, a quantity of steam, and merely allow it to expand, giving out heat in the process and doing work, you have no right whatever to say that the quantity of heat which has disappeared is the equivalent of the work which you have got, because at the end of the operation the steam is in a different state as to pressure and temperature from that in which it was at the beginning. It was saturated steam at a certain temperature, let us say, to start with, but at the end of the operation it may still, if you make proper adjustments, be saturated steam, but it is necessarily at a different temperature, and therefore you cannot tell whether or not it possesses intrinsically the same amount of energy as it did in its former state. You have no right whatever to reason upon the quantity of heat which appears to have gone, as compared with the work which has been done, when your working substance begins in one state and ends in another. But if you can by any process bring your working substance back to its initial state, then you are entitled to assert that, as it has returned to its initial state, it must contain neither more nor less energy than it did at first, and therefore of course you are also entitled to reason upon all the external things that have taken place during the operation, and to determine the condition of equivalence among them. You now see how completely unscientific was Séguin's reasoning, though his work was published fifteen years after that of Carnot. A similar remark of course applies to Mayer, who was the greater, because the later, sinner in this matter.

The other grand point with reference to Carnot is this,—that he started the notion of a *Reversible Engine*,—reversible not in the ordinary technical sense of working its parts backwards, not in the mere sense of backing, but reversible in the sense that, instead of using heat and getting work from it, you can drive your engine through your cycle the other way round, and by taking in work, pump back heat (as it were) from the condenser to the boiler again,—a reversing of the whole process,—not a mere reversing of the direction in which the engine is driving. Now, Carnot introduced that notion, and he showed by perfectly conclusive reasoning that if you can obtain a reversible engine, it is *the perfect engine*, *i.e.* that it is impossible to get an engine more perfect than a reversible one—reversible being taken in the sense in which I have just explained it. We see at once what an enormous step is gained, supposing we can establish that second principle, because, as you will presently find, we can settle the conditions of reversibility altogether independently of the nature of the working substance in our engine. You see then that we are not now bound down to a steam-engine, or any one working substance. We are enabled now to state our conclusions in terms, not of the particular engine but, of the circumstances in which the engine works. All perfect engines—that is, all reversible engines—will do exactly the same amount of work with the same amount of heat, provided their boilers and their condensers be at the same temperatures, and therefore you can define the relation between the whole amount of heat which enters the engine and the utmost amount of it which can be converted into work, and this altogether independently of the particular engine, but solely and simply in terms of the temperature of the boiler and the temperature of the condenser. These, then, are the grand claims which Carnot has in Thermodynamic Science.

Now, in order to make it intelligible how we can have a reversible engine at all, in this sense, it will be necessary for me to go through a series of imaginary operations explanatory of the nature of Carnot's reasoning. Besides, if you once thoroughly understand this, it gives the key to an enormous number of new physical facts and properties of matter which, before we learned from Carnot the correct method of reasoning, we might well have despaired of ever being able to understand, at least in their true physical interdependency.

Digression. Beam of ice, supported horizontally at the ends, with a fine wire, stretched by weights, hung over it.—Before I go into a description of it, however, I may call your attention to an experiment which has been going on for some time in your presence, and whose result, in one of its many forms, was first predicted from those very principles of Carnot's. What is its direct connection with them I shall explain in another lecture. In the meantime, the experiment is nothing more than this:—Take a block or bar of ice, supported horizontally: lay over it a fine wire, and append equal weights to the two ends of the wire. The wire, as you notice, has gradually, by the action of the weights, sliced through the bar of ice, and there are two such slices of which you can see the planes through the slab by the distortion of the air bubbles. The wire has actually passed through the ice in two planes parallel to one another, and yet the ice is now probably stronger at these two places where it has been cut than at any other place throughout the block. The statement of observed fact is, that as the wire was forced by the weights into the ice, the pressure melted the ice, making it colder, so that the water produced, passing round the chilled wire, and being thus relieved from pressure, froze again. Still the ice goes on melting in front of the wire, in consequence of the pressure, and the water formed continually trickles round it and freezes again. In that way the ice-block is reunited, and you would see no trace whatever of this interruption of it were it not for the fact that this particular mass of ice was originally full of air bubbles, and some of these bubbles having been permitted to escape during the passage of the wire, have left a transparent stratum which shows you where each section has been cut. Ice, in fact, being a substance which melts under sufficient pressure, behaves absolutely like a viscous or plastic substance, for it melts (and contracts) wherever the pressure is sufficiently great, thereby handing on the pressure to another part, and in so doing becoming solid again in its new form. Thus Forbes' *Viscous Theory of Glacier-Motion*, propounded in 1843 as a statement of observed facts, is seen to be but the necessary consequence of a remarkable physical property of ice.

Now, come to the consideration of this method of Carnot's. I take an ideal engine, because that is quite sufficient for the purpose of our reasoning. If our reasoning be correct, it is only a question of greater complexity to apply it to an engine of a more elaborate character. Suppose then we have the cylinder of a steam-engine—we shall dispense with the boiler altogether, because we shall, for the sake of simplicity, always make the cylinder its own boiler. Let us have in the cylinder a small quantity of water, and the piston pressed down so as to be nearly in contact with it. Suppose, then, that our piston and the sides of our cylinder are absolutely impervious to heat. That is another thing we cannot realise, but it will have important bearings when we come to consider what are the conditions of the reversibility of an engine. We shall find in fact that any loss of heat by conduction through the sides of the cylinder is fatal to the reversibility of the engine; but for all that, in our theoretical reasoning we assume that the sides of the cylinder and the piston itself are perfect non-conductors of heat. We also assume that the bottom of the cylinder is a perfect conductor of heat. These of course are all suppositions which cannot be realised in practice, but they serve to give us a conceivable and extremely simple engine to theorise upon. Suppose, then, we have three stands, on any one of which I may place this cylinder. The first of them I call A, the second B, and the middle one C. Now, suppose A to be a body which has a certain defined temperature, S, which is to be the temperature of the boiler. This body A is supposed to be constantly supplied with heat, so as always to be kept up (whatever happens) to that particular temperature. Then, B, which is to be used as the condenser, is to be kept constantly at a definite temperature T, lower than the temperature, S, of A. The third body is to be used merely for the theory of the operation; it has really no effect itself. It is simply a non-conductor of heat; it is in fact a sort of second bottom to be put upon the cylinder when it is not placed either upon the boiler or the condenser. Now, we can commence our operations in any order with this apparatus. The way in which Carnot did it is perhaps not the simplest, but it is historically the more important. We will commence, then, by setting the whole of this apparatus upon the hot body. The effect of this, as the bottom of the cylinder is a perfect conductor, is that the hot body begins at once to part with heat to the water inside, under the piston. The water then rises to the temperature S, and steam begins to form above it. This steam is limited in quantity by the space which is afforded for it, and by the temperature of the body. When as much steam has been formed as is consistent with these conditions, it is called saturated steam corresponding to the temperature S. Now suppose that, when things are in that condition, we allow the steam to expand or the piston to rise (the atmospheric pressure above the piston being easily neutralised by a counterpoise, especially in an imaginary engine), we could employ it to raise weights or do work of some kind or other externally. As it rises notice what takes place. The temperature remains the same as before, but more space is afforded for the formation of steam, and therefore more steam is formed, so that you go on keeping up saturated steam at the pressure corresponding to the temperature, S, of the boiler. As more steam is formed, more work is done, and more heat is absorbed from the boiler, because latent heat is required for the new steam as it is formed. Then, while things are in that condition—the piston having risen say midway up the cylinder—put the whole upon the body C. No heat can get into the cylinder now, nor can any escape, for the contents are now completely surrounded by non-conducting bodies. In that state, however, the contents have still the temperature of the boiler. Let them still further expand, they will still do work, because fresh steam is formed, but the contents will become colder because of the latent heat required. Let them go on expanding and doing work until they cool down to the temperature, T, of the condenser, and then, while they are in that state, shift the whole to the condenser. There will obviously be no transference of heat. While things are in that condition, suppose we spend work in forcing down the piston a certain way. In doing so we compress the steam, and the contents tend to become hotter, but cannot do so, because this body of temperature T is in contact with them; so that part of the steam condenses, and the latent heat which it gives out is transferred to the cold body.

With regard to the amount by which you must push down the piston during this part of the operation, Carnot said:—Push it so far that you give out to the condenser exactly the same amount of heat as you had taken from the boiler during the first stage of the expansion. That statement, however, is incorrect, and requires modification, because Carnot argued on the assumption that heat is indestructible.

Bearing in mind Carnot's notion of a cycle, we see that the amount by which the piston is to be depressed while the whole stands on the condenser, is to be determined by the condition that when the whole is finally placed on the impervious stand, and the piston pressed home, the temperature of the contents shall be raised to S, the temperature of the boiler. [This complete rectification of Carnot's cycle was given by James Thomson in 1849.] If this be effected, we can transfer the cylinder to the body A, and everything is in the condition from which we started, so that the operation may be repeated as often as we please.

LECTURE V.

TRANSFORMATION OF HEAT INTO WORK.

Carnot's Cycle—continued. Watt's Diagram of Energy. The Impossibility of *the* Perpetual Motion is an experimental truth. Conditions of Reversibility. Absolute definition of Temperature. Second Law of Thermodynamics. Absolute zero of temperature, or temperature of a body devoid of heat. Efficiency of the best steam-engine. Effect of pressure on the freezing point of water. Mechanism of Glacier motion.

You will remember that at the close of my last lecture I had just given a sketch of the first part of the reasoning of Carnot—the most important reasoning that has ever been introduced into the treatment of any part of the dynamical theory of heat. I may briefly recapitulate (but in a somewhat improved form) what I then said, in order that there may be no break of continuity.

The nature of the hypothetical operation which Carnot introduced for the purpose of reasoning on this subject, and only for that purpose, is of this kind. He said—Let us have a hot body which is constantly maintained at a certain temperature. Let us have a cold body which is also constantly maintained at a definite temperature lower than the first. Then let us suppose that in addition to these we have a body which, as regards other bodies, is neither cold nor hot, for the simple reason that it is incapable of absorbing heat or of giving it out,—a body which is a non-conductor of heat. Then commence your series of operations, not as I did (after Carnot) in my last lecture, but with the non-conductor. Suppose your cylinder and your piston to be non-conductors, but the bottom of the cylinder a perfect conductor. If you have a quantity of water and steam in the cylinder, both at the temperature of the cold body, and expend work in pressing down the piston, the contents will become warmer, and some steam will be liquefied.[1] Continue this process till the temperature rises to that of the hot body—then transfer the cylinder to it. Now allow the piston to rise, the contents remaining at the temperature of the hot body, fresh steam is generated, and work is done. Arrest this process at *any* stage and transfer the cylinder to the non-conducting body. If we now allow the contents further to expand, more work is done, but the temperature gradually sinks. Continue this till the temperature falls to that of the cold body, to which, therefore, without loss or gain of heat, it may now be transferred. Next apply work to compress it at the constant temperature of the cold body till (by condensation) the contents have become exactly as they were at starting. The cylinder may now be transferred to the non-conducting stand, and everything is as it was at first—save that some heat was taken from the hot body in the second operation, and heat was given to the cold body during the fourth. Also it is evident that *more* work has been done during the second and third operations than was spent in the first and fourth, for the temperature, and therefore the pressure, of the contents were

---

[1] [*Note to Third Edition.* This statement requires limitation, in order to avoid complications not alluded to in the text. If there be too small a quantity of water, as compared with the steam, pressure will vaporise some of the water, instead of, as is assumed in the text, condensing some of the steam. See Tait's HEAT, § 391.]

greater during the expansion than during the compression. Of course you can go over this operation as many times as you please.

Notice particularly what the peculiarity of the operation is. You must always have the steam or expanding substance, whatever it is,—for air or anything else would do equally well,—in contact with bodies at its own temperature, or else with non-conducting bodies. If it were in contact with a body which was not at its own temperature, there would be a waste of heat. Heat would pass by conduction from the cylinder to external bodies, and would of course be wasted as regards work. The same would happen if we were to take it from, let us say, the non-conducting body and place it upon the cold body, before we had let it expand far enough to cool down to the temperature of the cold body:—we should have some heat conducted away at once without having any good from it. So, throughout the whole of Carnot's operation, it is essential that there should be no direct transfer of heat at all except while heat is being taken in from the hot body or given out to the cold body: the temperature of the contents of the cylinder being in each of these cases the same as that of the body with which they are at the time in contact.

A remark of great importance must now be made, though it involves somewhat of a digression. You must have noticed how much more easily we managed in to-day's than in yesterday's lecture to lay down the limits for the range of volume of the working substance during each of the four operations included in Carnot's cycle. Yet the only difference in our proceedings consisted in the fact that yesterday, following Carnot himself, we began with expansion at the higher temperature

—while to-day we have preferred to commence with compression on the non-conducting stand. With the help of a device due to Watt it may be possible to make this point much more easily intelligible. The device I allude to is called the *Indicator Diagram*, and is even now constantly employed for the purpose of ascertaining the work actually done by an engine, especially that of a steam-ship.

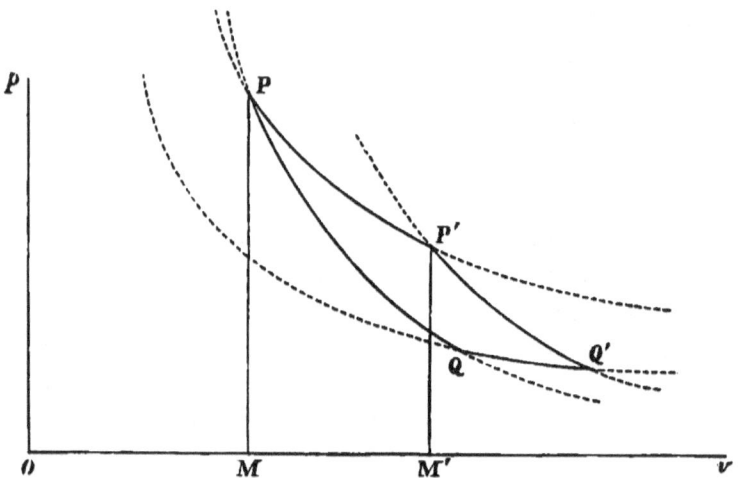

It is not my business to enter into purely mechanical details, and therefore I shall only say that this diagram is traced out by a pencil attached to the piston-rod of the engine, and therefore sharing its to-and-fro motion; while it has also a motion in a direction perpendicular to the piston-rod, such that the displacement at any instant is proportional to the pressure in the cylinder at that instant. To fix the ideas, suppose the cylinder to be horizontal, and the just-mentioned transverse motion vertical. Then any re-entrant line whatever (lying wholly between Op and Ov) may be supposed to be traced, once over in each cycle of the engine, by the pencil P. For reasons to be afterwards explained, I take the curvilinear quadrilateral $PP'Q'Q$. Let Ov be the axis of the cylinder, Op perpendicular to it; and let PM be perpendicular to Ov. Then, by our conditions, OM represents the distance of the piston from the bottom of the cylinder; i.e. *the volume of the working substance*, while MP represents its *pressure*, each upon a definite scale. It follows from this, by a mathematical investigation which, though very simple, I must not give in such a lecture as this, that if P' be any other position of the pencil, and $P'M'$ be perpendicular to Ov, *the area of the figure $PP'M'M$ is proportional to the amount of work done by the expanding substance* while the pencil passes from P to P'. Hence you easily see that the *area of the figure $PP'Q'Q$ is the excess of the work done by, over that spent on, the working substance;* i.e. the equivalent of the heat which disappears during the cycle.

Now, by properly regulating the temperature during the cycle, it is obvious that we may make the pressure what we please at each stage of the expansion and contraction. Hence any closed curve whatever might, by proper arrangement, be made the diagram of energy for a heat-engine. But now note particularly that in Carnot's ideal engine we are carefully restricted to two kinds of operations (direct or reversed), and to two only. Hence the parts of the indicator-curve for each of the four operations in Carnot's cycle belong to two classes of curves, each of which is known, or at least can be experimentally determined, so soon as we know what is the working substance.

One of these is the curve representing the relation of

pressure to volume when the working substance expands or contracts without change of temperature. Call this a line of *constant temperature—PP'* or *QQ'* in the diagram.

The other, *PQ* or *P'Q'*, represents the corresponding relation when the substance expands or contracts in a vessel impervious to heat. This is called, after Rankine, an *Adiabatic Line*. We might conceive Watt's graphical process actually applied to trace out these curves. And it is obvious that we can have *one*, and *only one*, of each kind, passing through each given point *P* in the plane of the indicator diagram. For that point specifies a particular volume and pressure of the working substance (treated as constant in quantity), from which we are to start by one or other of the two processes I have just mentioned. Also it is obvious that as, in general, the pressure of the working substance will fall off faster as it expands when no heat is communicated to it, than when its temperature is kept constant,—of the two lines passing through the point *P*, that corresponding to constant temperature *PP'* tends less quickly to fall to the line of no pressure *Ov*, than does the adiabatic line *PQ*, for equal increments of volume of the working substance.

You now see that Carnot's process essentially involves a cycle whose boundary (in Watt's diagram) is formed by two lines of equal temperature and two adiabatic lines. But, while these lines of equal temperature were at once specified by the numbers S and T, we had no such definite nomenclature for the adiabatic lines. Hence (so far as an elementary lecture is concerned) the greater simplicity of the method I have to-day used over that originally given by Carnot. To-day's method in fact began by taking any point Q in the line of temperature T, thence P was found by an adiabatic line: then P' may be any point in the other given line of equal temperature, and from this the adiabatic gives Q'. The difficulty in yesterday's method arose in specifying Q in the third operation, so that we should arrive at a *given* point P, in the fourth.

We now come to another point, also perfectly novel, and of importance at least proportional to its novelty. If you think again of the various steps of the operations in Carnot's cycle, you will easily see that it is possible to consider them as performed in exactly the reverse order. Begin, for instance, with the hot body, but do not allow the piston to rise there. Take the cylinder from the hot body when the water and the little quantity of steam above it have acquired the higher temperature. Lift it to the non-conducting body, and then allow the piston to rise. Let it rise till the temperature sinks to that of the cold body; place the whole on the cold body; allow it to expand still further,—it will be in that case giving out work but absorbing heat: then when it has risen to its former highest point, place it back again on the non-conducting body, force the piston back to the same extent as that to which it rose when (in Carnot's direct set of operations) it was first placed on that body. Everything has taken place in precisely the reverse order to that in which it took place before. Finish then upon the hot body, and press home. From Carnot's point of view you give to the hot body in that final operation precisely the quantity of heat you took from the cold body; but during the two last operations you are forcing down the piston, while during the two first operations the piston was being forced up, but it was always being forced up at a lower temperature, and therefore at a lower pressure than the temperature and pressure you had to overcome in forcing it home again. And, therefore, in the reverse method of working this engine, you take heat from the cold body and deposit it in the hot body, exactly to the same amount as in the direct operation; and, on the whole, you now *spend* as much work as you formerly *gained*.

These are the grand ideas which Carnot introduced. Their two distinctive features are, *first*, the idea of a complete cycle of operations, at the end of which the working substance, whatever it is, is brought back to precisely its primary condition; a cycle which can be repeated over and over again indefinitely. *Secondly*, The notion of making the cycle a reversible one, so that you can perform all the operations in it in the reverse order, and instead of taking in heat at any place it may be made to give out that amount of heat,—instead of the engine doing work at any place that amount of work can be spent upon it. With these changes in each operation, the whole cycle can be gone over the reverse way.

Now, Carnot proceeds to reason upon this. Considering heat as a material substance, he says that obviously it has done work in the direct series of operations by being let down from the higher temperature to the lower, just as water does work by being let down through a turbine or other water-engine, in proportion to the quantity that comes down and the height through which it is allowed to descend. We now know that this notion of the nature of heat is erroneous,[1] but still

[1] Carnot, as is now ascertained, had long ago found this out for himself. See note to p. 98.

Carnot's reasoning is of the highest value, because it wants only the change of a word or two to render it perfectly applicable to our modern knowledge of the subject.

You see at a glance one point which appears conclusively to show that Carnot's assumption was wrong, because nothing is easier than to let the heat down at once without the performance of any work. If you put the hot body into direct communication with the cold body, the same quantity of heat might be allowed to go down from one to the other, and yet give you no work at all. There must be, then, something wrong in that statement of Carnot. We now know what it is; but let us follow Carnot a little further, and see how much more of what is eminently useful and true he attained even with his false assumption. He carried it further in this way. He said—If an engine be reversible (as this cycle of operations has been shown to be), it does as much work as can be got out of a given quantity of heat under the same given circumstances. So that, no matter what you make your engine of,—no matter what be the substance which is expanding and contracting,— if a certain quantity of heat be let down from a source at a certain temperature through your reversible engine to a sink at another temperature, then the quantity of useful work which can be got from that heat will be absolutely the same. Reversibility is the sole necessary condition of equivalence between two such engines. You will see in an instant what an enormous step this is in physical science. The reasoning here is independent altogether of the properties of any particular substance. We are not dealing with steam, or air, or ether, or any one working substance in particular; yet we have a crucial test of the perfection of an engine which is absolutely the same when applied to any working substance and any heat-engine whatever. That test is, *if a heat-engine is reversible it is perfect*,—not perfect in the popular sense, but in a scientific sense; that is to say in the sense that *it is as good as it is possible physically to make it.*

Now the proof that it is so is very easily given, but before I give it I may say a word or two upon a similar but somewhat simpler sort of proof which will prepare you for the reasoning employed, and which bears directly upon the ordinary notion of the perpetual motion.

We know (of course *only* by experiment) that in all cases of natural laws, such as the laws of gravitation, and of magnetic attraction, whatever work is spent in moving a body through a certain course in one direction, you get back exactly by letting it return along the same track, always on the supposition that friction is avoided. The reason of this is that these forces depend upon relative position only, and therefore undo, at each stage of an exactly reversed path, precisely the amount of work which they did at the same stage of the direct path.

Suppose then that there could be two courses, from A to B, by the one of which more work would be spent on the mass than by the other. Let these amounts of work be W and w. I say that if such were the case you would be able at once to produce the perpetual motion. All you have to do is to apply frictionless constraint to guide the mass, so that in its ascent it shall travel along the course AwB, and in its descent along BWA. From A to B you have to spend the amount w of work against the forces of the system—from B to A these forces refund the amount W. On the whole, after a complete cycle, the mass is restored to A with an amount $W-w$ of energy additional to what it possessed at starting. Well, we have gained something by that, and every time the mass goes round the double course in the direction I have indicated, it gains the difference between the larger quantity and the smaller one, and therefore you can, at the end of each complete cycle, drain that amount off to turn some machine;—to do useful work. If, therefore, there were one way of doing a thing at less cost than another, and if the more costly operation were reversible (in the strict scientific sense above explained), then it would be possible for you under such circumstances to get unlimited amounts of useful work from nothing. Now we know that, so far as experience extends, this is impossible. The multiplied experiments of some of the most ingenious men who ever lived, Vaucanson and others, were directed to this question. Yet these men, who constructed automata which mimicked, and often copied, the motions and physical functions of living animals,—these men were entirely baffled in attempting to get at anything like the perpetual motion. We may say distinctly that all really scientific experiment has led to the conclusion that the perpetual motion in the old sense is absolutely unattainable.

Well, let us see how this reasoning applies to Carnot's engine. He demonstrates its property by almost the same application of reasoning as that which I have just given you for a similar but very simple case. He says that a reversible heat-engine is a perfect one; for, if not, let us suppose there could be one more perfect. Well, you can always use these two engines in conjunction. Let the more perfect engine (*i.e.* the *less* costly one) be employed in taking a quantity of heat, conveying it down to the condenser from the boiler, and giving you from it a larger quantity of work than the reversible engine could do. You can now use the reversible engine to pump that heat back again. Every time the heat goes down, it is through the more perfect engine; every time it is coming up, it is through the worse engine, and therefore it does more work going down than requires to be spent on bringing it up, and thus every time the compound engine makes a complete stroke, or passes through the double cycle of operations, you have an excess of work given by the one part over what has to be spent on the other. Therefore, this is not merely an engine which will go for ever, but an engine which can go on for ever, and besides steadily do work on external bodies.

That, however, as we have seen, is inconsistent with all our experimental results, and therefore we must at once pronounce the supposition which led us to this conclusion, viz.,—that there can be a more perfect engine than a reversible one,—to be false. This is Carnot's final proof that (on the assumption that heat is matter) a reversible heat-engine is a perfect engine. It requires very little indeed, as a moment's reflection will show you, to make this reasoning consistent with our modern knowledge of heat.

We have now to consider the cycle in the light of

the conservation of energy, so that if you get work from heat at all, some of that heat must have disappeared in its production, and that, therefore, under no circumstances—if the engine is doing external work at all—can the quantity of heat which reaches the condenser ever be equal to that which leaves the boiler. The difference between them,—if none has been wasted by conduction or in other unprofitable ways,—the difference between the quantity which leaves the source and the quantity which reaches the condenser during a complete cycle must be precisely the equivalent of the external work which has been done. Taking that into account, let us suppose we could make an engine more perfect than a reversible one. Work the two together, as before. Make the reversible engine continually pump up just as much as the other lets down. Then, as it is less perfect, it will require less work to be employed on it, when reversed, to restore to the source or boiler that quantity of heat than the other engine will do in letting it down; and therefore, on the whole, while you have a pumping up of heat and letting it down which will exactly compensate one another, or appear to do so, at least so far as the source is concerned, you will have a gain of work. There is the one point where the difficulty is to be found, if there is any. The compound engine will do work; no question of that. The more perfect engine lets down a certain quantity of heat to the condenser. The other engine pumps up heat from the condenser, and deposits in the boiler precisely the same quantity as the other takes out from it. How is it then, that, though we know heat is not matter, this double system can do work? It can only work in one possible way, and that is by expenditure of heat—it must ultimately work, therefore, not by letting down heat from the boiler, but by *cooling the condenser*. That is to say: If there can be a more perfect engine than a reversible one, then, with our present knowledge of heat, and taking Carnot's cycle, modified so as to make it compatible with our modern knowledge, these two engines, working together,—the one restoring to the boiler precisely what the other took from it,—can only do work, on the whole, on external bodies by cooling and further cooling the condenser. Hence, our result amounts to this, that by taking, as the condenser for our compound engine, any limited portion of the available universe, we could go on getting work from that by making it constantly colder and colder, till we removed all heat from it. Now, we may safely assume it to be axiomatic that we cannot do this; all experimental laws are against it; and as we see that the supposition that a more perfect engine than a reversible one can exist has led us to this absurdity, we have it *ex absurdo* that there can be no engine more perfect than a reversible one. What I have just given you is, in a much amplified form, the gist of some of Sir W. Thomson's remarks of 1851 on this point.

Clausius, in the preceding year, had endeavoured to supply this defect in Carnot's work by an appeal to the general behaviour of heat, *i.e.* its always striving to pass from a warmer body to a colder one. I have elsewhere given reasons which seem to show this proof to be inadmissible.[1]

However complete and satisfactory the demonstration just given may appear to be, you must now be told

---

[1] See the correspondence in full in the *Phil. Mag.* 1872, I. pp. 106, 338, 443, 516, and II. 117, 240. Also 1879, I. p. 344. The last of these is referred to in the *Preface*.

that it is possible, but possible only in a very curious way, and to an extremely limited extent, to get round this apparent difficulty,—to make a body colder than surrounding objects, and to get work from it in consequence. This (which, alone, is absolutely fatal to Clausius' reasoning, even with his later modification of it) was first pointed out by Clerk-Maxwell not long ago, and he showed that the mode of escape from the difficulty is, that it would require the intervention of beings, still finite, but infinitely more acute and able than any human beings (or even than the utmost ideal a human being can well conceive), to effect the object *on a finite scale*, and thus upset the basis on which Carnot's results have been re-established by Thomson. Clerk-Maxwell's reasoning depends upon the molecular theory of gases, an essential feature of which is that even in a mass of gas undisturbed by currents, and of uniform temperature, the particles have not all the same velocity. He points out that if such imaginary beings, whom Sir W. Thomson provisionally calls *demons*—small creatures without inertia, of extremely acute senses and intelligence, and marvellous agility—were to watch the particles of a gas contained in a vessel with a partition full of trap-doors, also devoid of inertia; prepared to open and close these doors so as to let the quicker particles get out of the first compartment into the second, and to let an equal number of the slower ones escape from the second compartment into the first, they could, without doing any work themselves, give to the system the power of doing a certain amount of work without help from external bodies. You must be careful, however, not to fancy that there is here any gain or creation of energy—not even a demon could effect *that*—there is a gain of transformability, a slight rise in the scale of availability—*voilà tout*. As you will be told in another lecture, this restoration of energy is constantly going on, but on a very limited scale, in every mass of gas. If there were only a few hundred particles in a small vessel of gas, the chances would be that if we suddenly cut off half the vessel there would be a sensible difference of temperature between the two parts. But, in consequence of the enormous number of particles in a cubic inch, *of even the most rarefied gas*, the amended form of Carnot's reasoning just given must be taken as holding good for every heat-engine. For, alas! we are not demons (in Maxwell's sense), and therefore, so far as experiment goes, and practical application goes, we may take this improved form of Carnot's demonstration as being absolutely decisive of the important result, that no heat-engine can be more perfect than a reversible one. This is, virtually, the *Second Law* of Thermodynamics, the *First Law* being that of the Equivalence of Heat and Work.

The consideration that follows immediately upon this is : If all reversible engines are perfect, they must all be of equal efficiency. They must all be able to give you precisely the same amount of work, from the same quantity of heat, under the same conditions. Therefore it follows that it is these CONDITIONS alone which determine how much work can be produced by a perfect engine, from a given quantity of heat. Now, what are the conditions? I have mentioned no conditions whatever but the temperatures of the boiler and condenser. The temperatures of the boiler and condenser were the only things this set of perfect engines had in common. Suppose they were all worked for such a period of time that they would all employ equal quantities of heat, then all would do the same amount of work. Therefore, having established Carnot's result, independently of Carnot's erroneous assumption, we are entitled to conclude that the perfect heat-engine converts into work a fraction of the heat it uses, the value of which fraction depends only upon the temperatures employed. Hence follows immediately one of the most important consequences of Carnot's method. It was given, as I have already said, by Sir William Thomson in 1848, when he first recalled attention to Carnot's work. He pointed out that here we have an *absolute* method of measuring temperature. All previous methods had depended on the properties of some particular substance. It is no matter what your zero and your Newtonian fixed points may be,—let us suppose them defined by melting ice and boiling water. Take a number of carefully made and calibrated thermometers; fill one with mercury, one with sulphuric acid, and so on, and one with water. All of these, if properly adjusted, will agree at the zero or freezing point and at the boiling point, but no two will, in general, agree at any intermediate point. In fact, the water thermometer would be an extremely curious thing, because for a few degrees from the freezing point upwards the water contracts instead of expanding. The water, heated from the freezing point, would at first go downwards on the scale, and then rise with increasing rapidity towards the boiling point. The mercury, on the other hand, would go pretty uniformly up, and so on. Thus, in employing such instruments you must, in addition to noting the degrees on the scale, also note the particular liquid employed. The temperature, then, of a body measured by thermometers filled with different substances, would give generally as many different readings as there are thermometers; and, therefore, unless you state what is the particular liquid employed, and even what is the particular kind of glass employed, your reader cannot be sure of the particular temperature which is meant. But Sir William Thomson pointed out that the reversible cycle gives us the means of defining temperature *absolutely;* that is, with complete independence of the properties of any particular substance, because Carnot's engine, if only reversible, is perfect. We do not need to inquire what is the working substance—air, water, chloroform, or ether, or whatever it is—the engines are all equivalent to one another, and the fraction of the heat they take in, which is converted into useful work, depends solely on the two temperatures. And we have seen that for a reversible engine it is only necessary that the working substance should never be in contact with a body of a temperature different from its own, unless indeed it be an absolute non-conductor of heat. Now, suppose we could keep a body at the temperature of boiling water, under certain conditions, such as that the barometer shall be at a height of 760 m.m., or, roughly, 30 inches. Suppose we keep another body at the temperature of melting ice, with the barometer at the same height. Suppose we can measure what amount of heat is taken in and what amount given out by a perfect engine working between these two temperatures, we should find that they are as nearly as possible in the proportion of 374 to 274. I make this statement just now simply as an assertion; we shall see afterwards by what process these numbers have been approximately obtained. In the ordinary Centigrade scale we call the freezing temperature zero, and we call the temperature of boiling water, under the 30 inches of pressure of the atmosphere, 100°. The experimental numbers have been so taken that their difference is 100, for a reason which will immediately appear. It is obvious now that we may define the temperatures of the boiler and condenser of a perfect engine by any functions of the relative quantities of heat taken in and given out. Sir William Thomson's first suggestion was not that which he finally adopted. To give as slight a dislocation as possible from the common modes of measuring temperature, it was found best, as it is also simplest, to define as follows :—When a perfect engine takes in heat at one temperature and gives it out at another temperature, then the temperatures of the source and of the refrigerator are in proportion to the quantities of heat taken in and given out, so that, as we see by experiment in the case above mentioned, that for 374 taken in, 274 are given out, the temperature of boiling water will on this scale be represented by 374°, and of freezing water or melting ice by 274°, the range between these being the ordinary 100° of the Centigrade thermometer. Therefore we have this curious result, that if you could get a body cooled down far enough under the freezing point—we have many artificial processes for such cooling: we can go nearly 140 degrees Centigrade below the freezing point,—if we could go twice as far, or 274 degrees below zero, we should have taken all the heat out of the body, we should have reduced it to the absolute zero of temperature. It would be impossible to make it any colder than the absolute zero of temperature just stated as 274° C. under the freezing point of water. Otherwise an engine could be constructed which would give more work from a quantity of heat than its dynamical equivalent. And this engine would work by taking heat from a body already more than totally deprived of heat!

In passing, I may say a word or two illustrative, perhaps even to be regarded as corroborative, of that conclusion. It has been long known that the pressure of a gas, such as air, in a closed vessel, becomes greater as you make it hotter. Take a vessel enclosing a quantity of gas, and shut off the connection between the interior and the exterior, and then apply heat to it. We know that the gas presses more strongly on the containing vessel. On the other hand, if, instead of applying heat to it, we immerse the vessel in a freezing mixture, we know that the pressure becomes less. Now, the amount of increase of pressure per degree of increase of temperature, and also the amount of diminution of pressure per degree of diminution of temperature, have been carefully measured, and it has been found that almost exactly—not quite exactly, for a reason afterwards to be assigned, but quite nearly enough for our present purpose—if you were to suppose the gas cooled down to a temperature of 274° C. under freezing point, and calculate, by assuming the experimental results I have mentioned to hold throughout that whole range of temperature, the pressure thus deduced would be almost exactly nothing. So that on the assumption that the formula for its dilatation (found experimentally for small ranges of temperature) holds for great ranges also, a gas would cease to exert any pressure upon its containing vessel if you could cool it down to 274° under ordinary freezing point, the degrees between the freezing and boiling points being, as in the Centigrade scale, 100. This, taken in connection with Carnot's result, appears conclusively to show that the pressure of a gas is due to motion of its particles. The application of heat produces this motion of its particles, in virtue of which they fly about and impinge upon the walls of the vessel; the energy of their motion is the heat contained by the gas. Go on cooling, and their motion becomes slower; and finally, when you have got the gas to the absolute zero of temperature, their motion will have ceased, and therefore we should find no pressure upon the retaining vessel.

I may now mention, in connection with the production of work from heat, and as a practical illustration of it, that suppose we could get a steam-engine made to fulfil Carnot's condition of reversibility—that is to say, that we could prevent the steam from ever being in contact with bodies at other than its own temperature for the time being,—prevent loss by conduction, etc.,—in other words, suppose we had a perfect engine, the fraction of the whole heat employed which would be converted into work would not be a large one. Using data, which I take from a statement of Joule, as fairly representing a practical case; suppose the engine to be a high-pressure one, working at $3\frac{1}{2}$ atmospheres, or something about 53 lbs. pressure on the square inch, it would require in the boiler a temperature of very nearly 300° Fahrenheit. Joule says that while working such an engine at an ordinary rate of speed it is next to impossible to keep the condenser colder than about 110° Fahrenheit. Now the question is, what fraction of the heat spent upon that engine would be converted into useful work? The answer is—remembering Carnot's cycle again—the quantity of heat taken in is to the quantity given out in the proportion of the absolute temperature of the boiler to the absolute temperature of the condenser; so it comes to be a question simply of arithmetic. Two hundred and seventy-four degrees under zero Centigrade is the point of absolute cold; what corresponds to that upon Fahrenheit's scale? This is easily found to be 461°·2 under the Fahrenheit zero. And, therefore, 761°·2 is the absolute temperature of the boiler; and 571°·2 will be the absolute temperature of the condenser. Therefore of 761·2 units of heat which go in, only 571·2 units go out; and as the engine is perfect, all the rest, that is, 190 units, amounting almost exactly to one-fourth, is converted into work. So our engine, under these conditions, which are about as favourable as any occurring in practice, and even with the additional assumption that it is a perfect engine—a thing quite unrealisable in practice—converts only one-fourth of the heat from the boiler into useful work. The other three-fourths are sent to the condenser, and in general wholly and absolutely wasted.

I come now to the consideration of various important advances in the pure science of natural philosophy, which have been made possible, or have at all events been brought forward sooner than they otherwise would have been, in consequence of the recognition of this great discovery of Carnot. One of the first of these, and certainly one of the most important, is that made by James Thomson, with regard to the effect of pressure upon the freezing point of water. As you will find immediately, the whole effect is, even for great pressures, an extremely small one,—and yet, in all probability it has fitted ice to be one of the most important agents in modifying physical geography.

Let us consider for a moment that when water freezes there is very considerable expansion. With a very slight change of temperature of water near the freezing point you have a very considerable change of bulk. Taking Carnot's engine again : Suppose that instead of putting into our cylinder a quantity of hot water with a little steam above it, we put a quantity of cold water with some ice in it, which went through the same set of operations ; then—and this was almost precisely the way in which James Thomson regarded it—it is easy to show that, taking account of the expansion in the act of freezing, you could get, without any expenditure of heat whatever, any amount of work you pleased from such an ice-water engine. The only way in which you could get out of this inadmissible difficulty is by assuming that the freezing point of water depends upon the pressure. If this be allowed, everything can be explained ; but if not, then unquestionably an ice-engine would enable us to get work from no expenditure. Thus, by simply applying Carnot's process with the change of a word or two, and availing himself of the experimentally demonstrated impossibility of the perpetual motion, James Thomson made out the result, that the freezing point of water is necessarily lowered by pressure. Well, one can calculate, suppose it were not lowered, how much work could be done in one stroke of this compound engine. One can compare that with the work done by expansion of water when converted into ice and the amount of latent heat set free, and from these one can calculate conversely by how much the pressure must be increased in order to lower the freezing point one degree, or how much the freezing point would be altered by a change of one additional atmosphere of pressure. Thomson made both these calculations. The result was extremely small, namely this fraction—

0°·0075 C. The freezing point of water is lower by this small fraction of a degree Centigrade for every additional atmosphere of pressure. You can calculate from this that it would require 133 additional atmospheres of pressure, that is to say, 133 times 15 lbs., or about 2000 lbs. weight on the square inch, to be applied to a quantity of ice which has a temperature one degree Centigrade lower than the freezing point, in order that the ice should melt. So that ice can always be melted by applying pressure great enough ; but if you make the ice very much colder than the freezing point, the amount of pressure required to melt it is so great that we can hardly conceive of its ever being applied. It is only when ice is moderately near its melting point that you can apply sufficient pressure to get its present temperature to represent its melting point; and if its present temperature represents its melting point, of course it melts. I showed you in my last lecture one beautiful method of exhibiting the melting of ice under pressure, which was described last year in *Nature* by Mr. Bottomley. It consisted in cutting through a bar of ice with a wire, as you would cut soap or cheese. But the ice behaves in a totally different way from that in which soap or cheese would have behaved under the same circumstances. If the ice had been one or two degrees colder than the freezing point, the wire would have hung inactive. It is only when the ice is near the freezing point that the wire, with moderate weights at its ends, is capable of melting it. As the ice melts, it passes round behind the wire, and, thus escaping the pressure, sets into ice again. Thus, as fresh ice has pressure applied to it by the advancing wire, there is a constant melting of the ice before the wire, and a constant re-freezing behind it; and the block of ice remains practically continuous, except just at the place where the wire is cutting it. Now, this property of ice was known in some of its effects—at all events to every one who had seen a glacier—for hundreds of years; but it was only within comparatively recent times that attention was directly called to it. The first who seems to have done so was Dollfuss-Ausset, in his experiments upon the Swiss glaciers, where he showed that by compressing a number of fragments of ice in a Bramah press, it was possible to melt them; and when pressure was taken off them, to allow them to revert again into a solid block. But he found that with very cold ice the experiment did not succeed. In fact, as we now see, even with his Bramah press, he could not apply pressure enough. Another form in which it must have been well known for hundreds of years is the form in which we try the same experiment every time we make a snowball. Schoolboys know well that after a very frosty night the snow will not 'make:'—their hands cannot apply sufficient pressure. But if the snow be held long enough in the hands to be warmed nearly to its melting point it recovers the power of 'making,' or rather of 'being made.' Every time we see a wheel-track in snow we see the snow is crushed, and even after one loaded cart has passed over it,—certainly after two or three have passed, —the snow has been crushed into clear transparent ice. The same thing takes place by degrees after people enough have walked over a snow-covered pavement; and in all these cases this minute lowering of the freezing point has led to the result. And now we see how it is that the enormous mass of a glacier moves slowly on like a viscous body, because in consequence of this most extraordinary property it behaves under great pressure precisely as if it were a viscous body. The pressure down the mass of a glacier must of course be very great, and as the mass is—especially in summer—freely percolated through by water, its temperature can never (except on special occasions, and then near the free surface) fall notably below the freezing point. Now, in the motion of the mass on its journey, there will be at every instant places at which the pressure is greatest,—where in fact a viscous body, if it were placed in the position of the glacier ice, would give way. The ice, however, has no such power of yielding; but it has what produces quite a similar result—wherever there is concentration of pressure at one particular place it melts, and as water occupies less bulk than the ice from which it is formed, there is immediate relief, and the pressure is handed on to some other place or part of the mass. The water is thus relieved from the pressure by the yielding caused by its own diminution of bulk on melting. The pressure is handed on; but the water still remains colder than the freezing point, and therefore instantly becomes ice again. The only effect is that the glacier is melted for an instant at the place where there is the greatest pressure, and gives way there precisely as a viscous body would have done. But the instant it has given way and shifted off the pressure from itself it becomes ice again, and that process goes on continually throughout the whole mass; and thus it behaves, though for special reasons of its own, *precisely* as a viscous fluid would do under the same external circumstances.

LECTURE VI.

TRANSFORMATION OF ENERGY.

Further consequences of Carnot's ideas. Anomalous behaviour of water and of india-rubber. Application to rock masses, and the state of the earth's interior. Availability of energy, and loss of availability. To restore the availability of one portion of energy, another portion must be degraded. Dissipation of energy. Sources of Terrestrial and of Solar Energy. Energy of plants and animals. Measure of the Sun's Radiant Energy. Energy now in the Solar System.

I SHALL commence this afternoon by taking a few further consequences of the grand ideas of Carnot, which I developed at full length in my last lecture. Wherever, in fact, we meet with any one anomalous physical result, we almost invariably find that it is associated with other anomalous results; and perhaps it is in this respect that Carnot's ideas have been of the greatest use in giving us new information.

Let us take, for instance, what I incidentally mentioned in connection with thermometers in my last lecture,—the fact that water would be an exceedingly bad substance to employ for the purpose of filling a thermometer bulb, because, even supposing that it did not burst the bulb when it froze,—supposing that we were using it from zero of Centigrade scale up to 100°, it would begin to contract when first heated, and would continue to do so up to the temperature of 4° C., and then it would expand like most other liquids. Now, here is a substance, which, when heated, becomes less in bulk: it contracts instead of expanding. We should expect, therefore, to find that water exhibits some other anomaly,—really the same thing if we could understand exactly all about the physical question involved, but appearing very startling to us when presented as something apparently quite new and different.

Let us look closely into the circumstances of this question. We are applying heat to water, and in consequence the water is contracting instead of expanding. Suppose, then, that we take water at a temperature between zero and 4° C., and apply pressure to it, what should we expect? Pressure applied to water at any temperature above 4° C., and to most other liquids at any temperature whatever, develops heat. Now Carnot's reasoning shows that just for the same reason that pressure produces a development of heat in a liquid which expands by heating, so in a liquid such as water between zero and 4° C., a liquid which contracts on being heated, pressure produces cooling, so that water taken at any temperature between these narrow limits and squeezed in a Bramah press becomes colder in consequence of the forced contraction in bulk.

Another very startling result is derived from the anomalous behaviour, which I daresay is familiar to most of you, of an india-rubber band. I daresay you all know that an india-rubber band suddenly stretched and applied to the lip feels warmer than before. Most bodies when extended become colder, as air does when it expands. If you pull out a steel spring it becomes colder, as Joule showed by direct experiment; but india-rubber is an exception: it not only becomes warmer when it is pulled out, but if,—keeping it still pulled out—you allow it to cool to the temperature of the air, and then suddenly allow it to contract again, it is very much colder than before, as you feel by applying it again to your lip.

Now these other bodies, such as air and the steel spring, when heat is applied to them, expand. A steel spring supporting a weight, and with heat applied to it, will expand, and allow the weight to descend. On the contrary, as I hope to be able to show you by a simple arrangement, when you apply heat to stretched india-rubber, instead of expanding it contracts, and perfectly in accordance with the theoretical prediction of Sir William Thomson from Carnot's reasoning applied to this case.

I suppose the spot of light crossed by a sharp horizontal dark line is visible to all of you near the top of the scale. The light from an incandescent lime-ball passes through a lens, and (after reflection from a plane mirror) is brought to a focus on the scale. The horizontal dark line is the image of a wire stretched in front of the lime-ball. This is our index, not the vaguely-defined spot of light. I have here suspended a piece of vulcanised india-rubber gas-tubing, with the spiral wire-core removed from it. Its lower end has a scale-pan attached, and is also fastened to a lever which moves the plane mirror. In order to show you what the movements of the apparatus indicate, my assistant will put one or two additional weights into the scale-pan hanging from the tube, and you will notice that the effect of the additional weights, which is of course to extend the india-rubber, produces a movement of the reflected light *up* the scale. Hence, if this india-rubber were to expand further by the application of heat, we should see the spot of light on the scale move farther up; but, on the contrary, as soon as heat is applied by a spirit-lamp to the india-rubber, the spot of light you see moves *downwards* upon the scale, showing that the india-rubber is contracting instead of expanding. India-rubber is a very bad conductor of

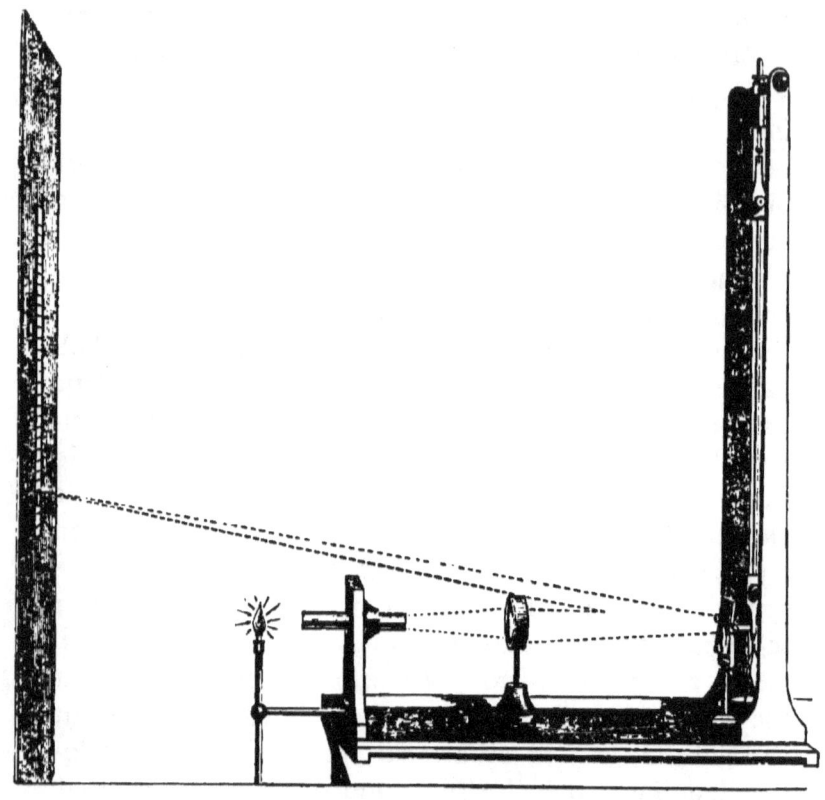

heat, so that it will require some time to cool; but if we were to allow it time to do so, we should find it return almost exactly to its original length; so that while being heated under tension it contracts, and while cooling under tension it expands.

[Clerk-Maxwell has recently improved this experiment in a most marked manner, by heating the india-rubber tube by the passage of a current of steam through it. The shortening produced can now easily be made visible *directly* to a large audience.]

There are a great many other substances which present anomalous properties of the same kind; but we will now go back to cases which are not anomalous, and there we shall see that the application of Carnot's principles leads, in these as in other cases, to results which may be of the very greatest importance. Take, for instance, a piece of wax. We know that when wax solidifies it contracts very considerably. Paraffin and many other bodies do the same; and, therefore, in exact accordance with Carnot's reasoning, their melting points are raised by pressure instead of being lowered, as the melting point of ice is, so that in order to melt paraffin under a very great pressure, you require to heat it very much above its ordinary melting point.

This is exactly analogous to the case of the conversion of water into steam. When water is converted into steam, there is an enormous increase in the bulk, and we know that the temperature of the boiling point of water is greatly heightened by increased pressure. In a high-pressure steam-engine, and wherever we insist on having steam at a high pressure, the boiler requires to be raised to a correspondingly high temperature above the ordinary boiling point. We all know that Papin's Digester was formed upon that principle, for the purpose of heating water to a very much higher temperature than the ordinary boiling point, and therefore to confer upon it solvent powers for dissolving bones and such like, which it does not possess at the ordinary boiling point. And, in the same way, Alpine travellers have told us of their difficulties in cooking tea and other food on the top of a high mountain, because it is impossible at such altitudes, without enclosing the water in a boiler with a closed lid, to heat it up to the temperature of 100° C., the ordinary boiling point. Water boils in an open vessel at about 85° C. on the top of Mont Blanc.

Now, consider the application of this on a far more gigantic scale. Think of its application to the (originally fluid) substances which now form the whole mass of the earth. There can be no doubt whatever, from various physical and geological proofs, that the interior of the earth, at all events for a very considerable depth under the surface, must, at some long time ago, have been in a viscous or even a perfectly liquid state. Now, when that mass first cooled, which it certainly would do most rapidly at the surface, then if the substance were such as to contract on cooling, so that the solid crust became denser than the liquid below it, there would be an exceedingly precarious state of equilibrium, as gradually the crust formed, and, shrinking in, increased the pressure on the liquid below, and thus produced a powerful horizontal tension throughout its own substance. In all probability the crust must have broken up by the surface-tension necessary to balance this internal pressure (as the tension of a soap-bubble balances the extra pressure due to the compressed air it contains), and tumbled in (and sunk) in pieces, and then solidification commenced on the fresh exposed liquid surface, and so on. But through the whole globe, there may be, at depths even of so little as 500 miles under the earth's surface, portions still left of the originally liquid mass at temperatures equivalent to a red heat; or (it may be) even a white heat—at temperatures at all events far above their melting point under ordinary pressures; and yet, as Sir William Thomson has shown by means of precession, and by other astronomical determinations, still solid. The whole mass of the earth is virtually solid; more rigid in fact than if it had been of glass throughout—very nearly as rigid as if it had been a solid mass of steel; still, I say there may be portions of the interior, even not so much as 500 miles under the surface, which are still at a white heat, and yet solid, because in consequence of the immense superincumbent pressure their melting points have been raised so high that even a white heat is insufficient to liquefy them.

The illustrations of this lecture have been mainly devoted to the law of transformation of energy from one form to another, and all the examples I have given have been simply applications of Carnot's great result, as modified slightly so as to make it agree with modern knowledge as to the nature of heat. But there are other reflections which we must make on the same subject, and especially with reference to the necessity in Carnot's process of a large portion, by far the greater portion, of the heat which even a perfect engine employs, being let down, without undergoing transformation, from a high temperature, where it has a great deal of available energy, to a lower temperature, where it has a less amount of available energy.

There is, of course, the same amount of energy in a given quantity of heat in whatever body and at whatever temperature you have it; for a quantity of heat, whatever its temperature, represents its equivalent of work.

But though there is a definite mechanical equivalent for so much heat, there are vast variations in its utility under different circumstances. If you have the heat in a very hot body, you can get a great deal of its value out of it. On the contrary, if you have it in a comparatively cold body, you can get very little out of it; and therefore we are led to speak of the availability of an amount of heat-energy. Availability of energy simply means its capability of being transformed into something more useful, *i.e.* of being raised higher in the scale of energy; and depends in the case of heat entirely upon the temperature at which we have it.

We have seen that even a perfect engine, when it is using heat, necessarily converts only a part of the heat into work. We get the full benefit of that part of the heat; but the remainder is not left in the boiler, but is degraded, is let down, through the range of temperature corresponding to that between the boiler and the condenser; and there, although even yet it is equivalent as much as ever to work, it cannot be converted into useful work; for, in order that such a conversion should take place, we must have a new engine working down to a temperature lower than that of the original condenser. Therefore this heat, although quite as high as the rest in its equivalent of mechanical energy, is not so useful, because we have not the means of transforming it. It has lost its standing, as it were; it has lost its availability; and thus there is a constant tendency, even with a perfect engine,—and we cannot get a perfect engine in any of our operations,—to a degradation of the greater part of the heat employed.

This leads us, then, to the consideration of why it is that such a degradation must take place. Perhaps the best way of studying such a question will be to take —as another illustration of the perfect engine, and Carnot's cycle—the case of compressed air, or some other such source of power which does not necessarily involve the direct application of heat.

The case of compressed air is a very instructive one, and at the same time a very simple one. It was first thoroughly worked out by Joule, and in this way. He took a strong vessel containing compressed air, and connected it with another equal vessel which was exhausted of air. These two vessels were immersed each in a tank of water. After the water in the tanks had been stirred carefully so as to bring everything to a perfectly uniform state of temperature, a stop-cock in the pipe connecting the two vessels was suddenly opened. The compressed air immediately began to rush violently into the empty vessel, and continued to do so till the pressure became the same in both; and the result was, as every one might have expected, that the vessel from which the air had been forcibly extruded fell in temperature in consequence of that operation. It had expended some of its energy on forcing the air into the other vessel. But that air, being violently forced into the other vessel, impinged against the sides of that vessel, and thus the energy with which it was forced in through the tap was again converted into heat. Thus the air which was forced into the vacuum became hotter than before, while the air which was left behind became colder than before. But, on stirring the water round these vessels, after the transmission of air had been completed, and the stop-cock closed, Joule found that the number of units of heat lost by the vessel and the water on the one side, was almost precisely equal to the quantity of additional heat which had been gained on the other side.

He then repeated the experiment,—putting instead of two tanks of water, each holding one of the two strong vessels, one larger tank also filled with water, with both vessels buried side by side in it,—then, on allowing part of the air to escape, as before, from the one into the other, and stirring till everything had acquired exactly the same temperature, he found that there was scarcely any measurable change in temperature.

These experimental methods, then, proved indisputably that the quantity of heat lost by the one part of the air was—at least as nearly as that kind of experiment enabled him to test it—equal to the quantity of heat gained by the other. Now, think of this for a moment, and you will see that the compressed air had at first a certain capability of doing work. You might have used it to drive a compressed-air engine, or you might have used it for propelling air-gun bullets, or anything of that sort; but in its final state, when it had expanded to double its original bulk, it had not anything like such an amount of available working power stored up in it as it had before. There was, therefore, dissipation of the energy, or of part of the energy, originally present; and yet, as you have seen, the apparatus and its contents had not lost any heat.

There was, on the whole, no heat lost, because what was lost to the one vessel was gained by the other. No heat was given out to external bodies, and no available work was done. The air was simply allowed to expand—to change its bulk—without driving out pistons, or doing anything by which it could convey work to external bodies. It had, therefore, at last precisely the same amount of energy as at first; and yet of that not nearly so much was available. The air had seized at once the chance given it of dissipating part of its energy, and did dissipate it, as far as was compatible with the circumstances of the arrangement.

Now the really curious point about this is, that in order to restore the lost availability to the energy of the air,—to get that air back into its former condition, so as to be capable of doing as much work as it was capable of doing at first,—it would be necessary to spend work upon it, pumping it back from the double vessel into the single one; but the amount of work which is so spent in pumping it back goes to heat the whole mass of air; and when you have expended work enough to force back the air into the first vessel from the second, you find that the amount of heat which is given out during that process—and which can be measured with great exactness—is almost precisely equivalent to the work which is spent in forcing the air back.

Thus to restore to the energy its former availability, you do not need to spend any energy, you have only to degrade some. You have spent work and got instead its less useful heat-equivalent. You must waste a certain amount of energy, or rather get a bad form of energy in place of it, in order to restore to the mass of air the availability of the energy which it possessed originally, and which had been allowed to be lost by gradual expansion.

I can illustrate this in another and very instructive way by taking an experiment belonging to the domain of electricity. The experiment is, I daresay, a well-enough known one, so far as the mere exhibition of an experiment goes, but its really important feature, its explanation as bearing upon the principles of energy, and especially upon Carnot's results, does not appear to be, at all events, very generally known. I have got here a couple of Leyden jars, and, contrary to the usual practice, their exterior and interior coatings are both insulated. The jars are supported upon varnished glass stems. Now, I am going to charge from the electric machine only one of those two jars. First of all, we shall charge and discharge it; and you will be enabled

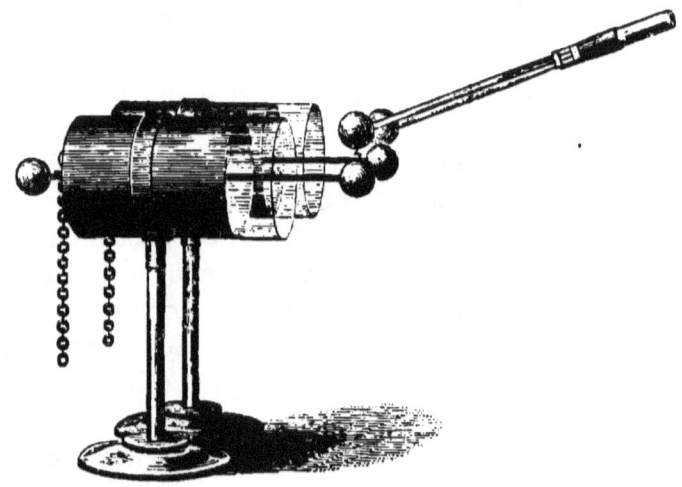

to judge roughly the amount of work which corresponds to its full charge by the sound and light of the spark. After that I shall charge it again as nearly as possible to the same amount, and then share the charge of electricity between the two jars, by putting first their outer coatings together, and then their inner coatings; so that the charge shall be divided equally (because of their equality) between the two. You now obtain (showing) from the sound and light of that discharge-spark an idea of the amount of work stored up in the jar when charged. Now, the jar being charged again, I simply place a chain over the two outer coatings, and then I connect the interior coatings by means of the discharging-rod. But you will notice that a spark passes during that process. (*Shows.*) Now, no electricity has disappeared, for the jars and discharging-rod are, all of them, insulated. But by separating the two jars from one another, and discharging them separately, you find there is a charge in each (*shows*), and that these are as nearly as possible equal, so far as can be judged by the appearance and sound of the discharges. But you must have noticed, also, that of the four sparks which you have just heard and seen, the first was very much the stronger; it made by far the greater noise, and it was also the longer and more brilliant. The second spark was the next in order of magnitude, and the two final sparks, as we should have expected, were about equal, but not at all comparable in intensity, even to the second one, which was weaker than the first.

Now, this is a beautiful illustration of exactly, or almost exactly, the same principle as that I have just explained. When I had the full charge of electricity in the one jar, there was a certain definite quantity of what, for want of precise knowledge, we provisionally call positive electricity, in the inner coating, and an equal quantity of negative electricity was in the outer coating. Then, when I connected the outer coatings of the charged and the uncharged jar by means of this chain, they formed, as it were, the outer coating of a single jar; but in order to make the two inner coatings correspond in electric condition, I had to put the discharging-rod between them, and you noticed that I could not do so without allowing a spark to pass. A spark *necessarily* passed during that operation, at least it did so when a short stout metallic discharging-rod was used.

Now, that spark represented a portion of energy which was wasted—a certain amount of work done in producing sound, light, and heat. Therefore, obviously, from the mere fact that such a spark passed when I completed the connection between the two jars, you saw that energy must have been wasted. But how could the energy be wasted when there was no free electricity lost? The quantity of positive electricity originally in the inside of one jar was simply divided between two jars;—there was just one-half the original quantity of positive and one-half the quantity of negative in each. The quantity of electricity remained the same, and yet there was a quantity of energy dissipated during the process. Now the answer is simply this (it was originally made out as a very particular case of grand general theorems, given first by Green and afterwards interpreted and applied by Helmholtz and Sir William Thomson),—that the work due to a charge of electricity, or the work which must be spent upon an electric machine (suppose it wholly goes to producing electrical charge of a conductor), depends upon the square of the quantity of electricity. No matter what the form of the conductor or jar is, the energy of the charge, or the amount of work which it will do, depends upon the square of the quantity of electricity. *Now* we can understand perfectly our experimental result. Suppose we call the quantity that the first jar had when it was charged, one; then, when I discharged it by itself on the first occasion, you had a spark which corresponded to the quantity of energy, the square of one, or one itself. But when I put the two jars together, and thus divided the charge, so that there was only one-half the quantity of positive, and one-half the quantity of negative in each jar, then the whole discharge of each separate jar, or the energy of it, was proportional to the square of one-half,—that is, to one-fourth. Each of these, when the charge had been divided between them, contained a quantity of energy equal to one-fourth of the original store, and therefore the two together corresponded only to one-half of that store. Now we can see what it was that produced the spark when I was dividing the charge: that spark was the equivalent of the other half of the energy, the half which necessarily went to waste. You wasted the whole quantity by discharging the charged jar itself; but in merely putting the two together, so as to divide the charge, you wasted one-half the energy, and then the quantities that you had remaining corresponded to the two remaining quarters.[1]

Now, in all these illustrations that I have shown you —whether they correspond to dissipation of ordinary energy, or to dissipation by sound or friction, or even to the production of heat, light, and so on, by electrical discharges,—in all these cases, you notice that there is a tendency for the useful energy, whenever a transformation takes place, to run down in the scale,—that, the quantity being unaltered, the quality becomes deteriorated, or the availability becomes less; and from similar

[1] If instead of the stout, short, discharging-rod I had used a very long, fine wire or other conductor of great resistance, such as, for instance, a number of persons joining hands, the second spark might have been reduced indefinitely; but then the inevitably wasted half of the energy would have appeared as heat in the wire, or in the physiological effects of the shock.

results in all branches of physics we are entitled to enuntiate, as Sir William Thomson did very early after the new ideas were brought into full development, the principle of Dissipation of Energy in nature.

The principle of dissipation, or degradation, as I should prefer to call it, is simply this, that as every operation going on in nature involves a transformation of energy, and every transformation involves a certain amount of degradation (degraded energy meaning energy less capable of being transformed than before), energy is continually becoming less and less transformable.

As long as there are changes going on in nature, the energy of the universe is getting lower and lower in the scale, and you can see at once what its ultimate form must be, so far at all events as our knowledge yet extends. Its ultimate form must be that of heat so diffused as to give all bodies the same temperature. Whether it be a high temperature or a low temperature does not matter, because whenever heat is so diffused as to produce uniformity of temperature, it is in a condition from which it cannot raise itself again. In order to get any work out of heat, it is absolutely necessary to have a hotter body and a colder one; but if all the energy in the universe be transformed into heat, and if it be all in bodies at the same temperature, then it is impossible—at all events by any process that we know of as yet—to raise the smallest part of that energy into a more available form.

Having seen then that this must be the ultimate end of all the energy in the universe; that—so long at all events as those I have just been explaining remain physical laws—this is the consequence to which they must lead, it becomes a very necessary inquiry—Whence is it that the enormous quantities of energy which are made use of, even on the surface of our diminutive planet, are supplied to us? What are our principal sources of energy, and how do we transform the supplies they afford us so as to make them useful for various practical purposes, especially the most practical of all, —the practical one of living, which, so far as mere vitality is concerned, is certainly a purely physical process?

Well, the muscular work which an animal does, and the animal heat which it gives out (in much larger equivalents than it does muscular work), these of course we all know are due mainly to food. In such a term as food, I include not merely solid and liquid food, but also (and this is very important) the gaseous food which we inhale. All these may be classed under the general title of food. These being taken in, we have certain other things which are got rid of, such as carbonic acid, water, and so on. These you may call the ashes of our food. These have, in their chemical relations, part of the degraded energy of the food which was taken in. The non-degraded part of the energy corresponds of course to the muscular work done, and the store of muscle, etc., laid up in the system.

Now, if this process were going on continuously there would be constant using up of the oxygen of the atmosphere by its combination with part of the food, and production of the (to the animal) useless, or rather pernicious, gas, carbonic acid. Leave this part of the question as a difficulty for the moment,—that we should have the oxygen of the air gradually taken up, and its place supplied, at all events to a great extent, with carbonic acid gas, in which an animal could not live:—still we have this further difficulty:—Although animals may live to a great extent upon animal food, yet if you go on from man, who consumes a certain kind of animal, while that animal also consumes animals, and so on, there must be either a cycle in which the last animal consumes man, or an infinite range as it were of animals, so that all could live on animal food!

We know that it is not so, that there is a large class of animals which consume only vegetable food. Now, it is to the wonderful difference between the application of the laws and processes of energy to the nutrition of animals, and to that of vegetables, that we are indebted for the explanation of the difficulty which I have just pointed out to you—what becomes of this large quantity of carbonic acid gas, which in time would, if not got rid of, kill off all animals, either by direct poisoning, or by depriving them of their oxygen. The explanation is simply this, that the animal takes in the oxygen, and with it animal or vegetable food, giving out the objectionable carbonic acid gas; but, on the other hand, the plant takes the carbonic acid gas, with water and other things, and works it up again,—gives back the oxygen to the air, and stores up the carbon, etc., in the form of vegetable food, upon which many animals live, and in their turn become man's food.

Now, it is quite obvious that if plants were not assisted by some external supply of energy, here would be something equivalent to the perpetual motion on the grandest conceivable scale. If the plant were capable, merely by its own peculiar organisation, of taking the ashes as it were of the fuel burnt in the animal engine, and working them up again into fit and proper food, without external assistance, then that process might go on indefinitely,—the animal all the time, remember, giving out animal heat and doing muscular work.

This would be the perpetual motion on a scale never contemplated even by the perpetual-motionists. It is obvious then that in order to escape from our difficulty —no less than a contradiction in terms of what we know to be a physical law—there must be some source of energy which the plant draws upon in order to help it to work up that carbonic acid, etc., and store up the available part of it as food for the animal.

It was long ago recognised, but first, perhaps, in a nearly definite form, by Stephenson, that it was by energy supplied in a radiant form from the sun, that plants were enabled to decompose carbonic acid; and it is a very wonderful thing that those so-called actinic or chemical radiations from the sun, which are most effective in promoting the decomposition of carbonic acid by the leaves of plants, are the very rays which are most absorbed by the green leaves. The green leaves are particularly absorbent of them, and any of you may convince himself of the fact by comparing the photograph of a tree in full leaf with that of almost anything else. In fact, the photographs of foliage (at all events with the chemicals usually employed) are almost invariably exceedingly dull, even black, showing that the chemically active rays, except those which have been reflected from the surfaces of polished leaves, have been absorbed at once by the green leaves, and in this act have been performing their function of decomposing carbonic acid and water.

In fact, we may make a rough comparison—it is not by any means an exact one, but it is close enough to be sufficient for our present purpose—we may compare roughly the animal to the cell of a galvanic battery, where you have the virtual food supplied in the shape of zinc and dilute sulphuric acid ; and the cell, by means of the electric current it produces, driving an electromagnetic engine or producing heat in a wire, just as the animal produces muscular work or animal heat. On the other hand, you may roughly, with about the same degree of approximate accuracy, compare the plant to a cell in which energy, in the form of a current of electricity, furnished from an external source, is employed in decomposing water, let us say :—separating it into its oxygen and hydrogen, and producing that high form of potential energy which I exhibited to you experimentally in a former lecture ; so that fresh materials, as it were, for the battery cell are being actually separated, and getting their potential energy given back to them in the decomposing cell. That corresponds to the plant. You supply these materials again to the cell of the battery, and it again produces electric currents, and so on in succession.

But it is quite obvious that a process of that kind cannot go on without a supply of energy from without. The raising of energy from the lower form to the higher always requires external application of some fresh energy, which is itself degraded in the process. This idea originated with Joule at a very early period of his investigations; and he pointed out that not only does an animal much more nearly resemble in its functions an electro-magnetic engine than it resembles a steam-engine, but he also pointed out that it is a much more efficient engine,—that is to say, an animal, for the same amount of potential energy of food or fuel supplied to it (call it fuel, to compare it with the other engines), gives you a larger amount converted into work than any engine which we can construct physically.

To use the vernacular of engineers on the subject, the 'duty' of an animal engine is much larger than the duty of any other engine, steam, or electro-magnetic, or otherwise, which we can construct to employ fuel, —the duty simply meaning the percentage of the energy of the fuel supplied to the engine which it can convert into the useful or desired form. Carefully observe here that this does not necessarily hold true if we contemplate water-mills, etc., for there the energy supplied is in general of a higher order than that of food or fuel.

Now, from what I have said, you will see that the supply which the plant requires comes from the sun. That leads us then to the question—what is the source of the sun's energy? Now, when, with the view of answering that question, we make a few calculations, we find that they at once upset the first ideas that we are likely to form for ourselves on the subject. Of course, the old notion that the sun is a huge fire, or something of that kind, is one which will only occur to those thinking of the matter for the first time; but with our modern chemical knowledge, assisting the more ordinary physical reasoning which I have just given you, we are enabled to say, that, massive as the sun is, if its materials had consisted even of the very best materials for giving out heat by what we understand on the terrestrial surface as combustion, that enormous mass of some 400,000 miles in radius could have supplied us with only about 5000 years of its present radiation. A mass of coal of that size would have produced very much less than that amount of heat. Take (in mass equal to the sun's mass) the most energetic chemicals known to us, and in the proper proportion for giving the greatest amount of heat by actual chemical combination; and, so far as we yet know their properties, we cannot see the means of supplying the sun's present waste for even 5000 years.

Therefore, as we all know that geological facts, if there were no others, point to at least as high a radiation from the sun as the present, for at all events a few hundreds of thousands of years back,—perhaps, as we shall find later, even a few millions of years back,—and perhaps also indicate even a higher rate of radiation from the sun in old time than at present—it is quite obvious to you that the heat of the sun cannot possibly be supplied by any chemical process of which we have the slightest conception.

Now, if we can find, on the other hand, any physical explanation of this, consistent with our present knowledge, we are bound to take it and use it as far as we can, rather than say—This question is totally unanswerable unless there be chemical agencies at work in the sun of a far more powerful order than anything that we meet with on the earth's surface. If we can find a thoroughly intelligible source of heat, which, though depending upon a different physical cause from the usual one, combustion, is amply sufficient to have supplied the sun with such an amount of heat as to enable it to have radiated for perhaps the last hundred millions of years at the same rate as it is now radiating, then I say we are bound to try that hypothesis first, and argue upon it until we find it inconsistent with something known. And if we do not find it inconsistent with anything that is known, while we find it completely capable of explaining our difficulty, then it is not only philosophic to say that it is most probably the origin of the sun's energy, but we feel ourselves constrained to admit it. Newton long ago told us this obligation in his *Rules of Philosophising.*

The shortest and easiest way in which I can illustrate this simple though tremendously important step is by stating that if we were to take a mass of the most perfect combustibles which we know,—those combustibles which give the greatest amount of energy when burned together, and let it fall upon the sun merely from the earth's distance,—then the work done upon it by the sun's attraction during its fall would give it so large an amount of kinetic energy when it reached the sun's surface as to produce an impact which would represent 6000 times the amount of energy which could be produced by its mere burning. It is, in fact, capable of perfectly easy and simple demonstration,— that a mass which would produce the utmost known energy by burning, would give 6000 times more energy by a simple fall from the earth's distance upon the surface of the sun.

It appears, then, that until it is shown that there is, or has been, in the physical universe, at some time or other, a greater amount of kinetic energy than can be accounted for by the falling together of the masses which compose the sun and stars, our natural and only trustworthy mode of explaining the sun's heat at present, in time past, and for time to come, must be something closely analogous to, but not identical with, what was called the nebular hypothesis of Laplace—very eagerly accepted when it was first proposed—the hypothesis of the falling together (from widely scattered distribution in space) of the matter which now forms the various suns and planets. We find, by calculations in which there is no possibility of large error, that this hypothesis is thoroughly competent to explain 100 millions of years' solar radiation at the present rate, perhaps more; and it is capable of showing us how it is that the sun, for thousands of years together, can part with energy at the enormous rate at which it does still part with it, and yet not apparently cool by perhaps any measurable quantity.

Now, in confirmation of this it is well to state here, that not only is the hypothesis itself capable of explaining the amounts of energy which are in question, but also recent investigations, aided by the spectroscope, —of which I shall have a good deal to say in another lecture,—have shown us that there are gigantic nebular systems at great distances from our solar system, in the process of (physical) degradation in that very way, by the falling together of scattered masses, and with immense consequent developments of heat by impacts. What are called temporary stars form another splendid and still more striking instance of it, as where a star suddenly appears of the first magnitude, or even brighter than the first, outshining all the planets for a month or two at a time, and then, after a little time, becomes invisible in the most powerful telescope. Things of that kind are constantly occurring on a larger or smaller scale, and they can all be easily explained on this supposition of the impact of gravitating masses.

Now, holding that such may be the cause of the enormous amount of radiation from the sun, let us inquire what fraction of that whole radiation reaches our own little globe. We know what an enormous quantity of solar heat reaches the earth,—reaches even our own small corner of the earth. That is of course a very small part of what reaches the earth's whole surface; but still, if you recollect that the earth, as seen from the sun, appears very much less than the planet Jupiter, or even Mars, as seen by us,—that is, that it would present no visible disc to the naked eye, and that to an observer at such a distance as that of the sun it would require a telescope of some little magnifying power to show it as a disc at all,—considering also that the sun is radiating very nearly uniformly in all directions,—how much of the sun's entire radiations can reach this little speck at such a distance as ninety millions of miles? A circular disc of four thousand miles radius, at a distance of ninety-one million miles, appears to occupy less than one two-thousand-millionth part of the celestial sphere. You see, then, that the quantity of heat which the whole earth gets from the sun is of the order of something less than the two-thousand-millionth part of that which the sun gives out. Now, experiments have been made, and fairly satisfactory ones, to determine what amount of heat we do receive—what amount of energy does fall upon the earth's surface in a given time. Of course, they are interfered with to a considerable extent by absorption of the radiation as it passes through the various and varying constituents of the earth's atmosphere in each region of the globe; and therefore the most trustworthy experimental results have been such as were obtained at considerable elevations in balloons, or on the tops of very high mountains, where there is comparatively little absorption.

This instrument, the pyrheliometer, is constructed for the purpose of measuring the amount of radiation from the sun. It is made of silver polished on the

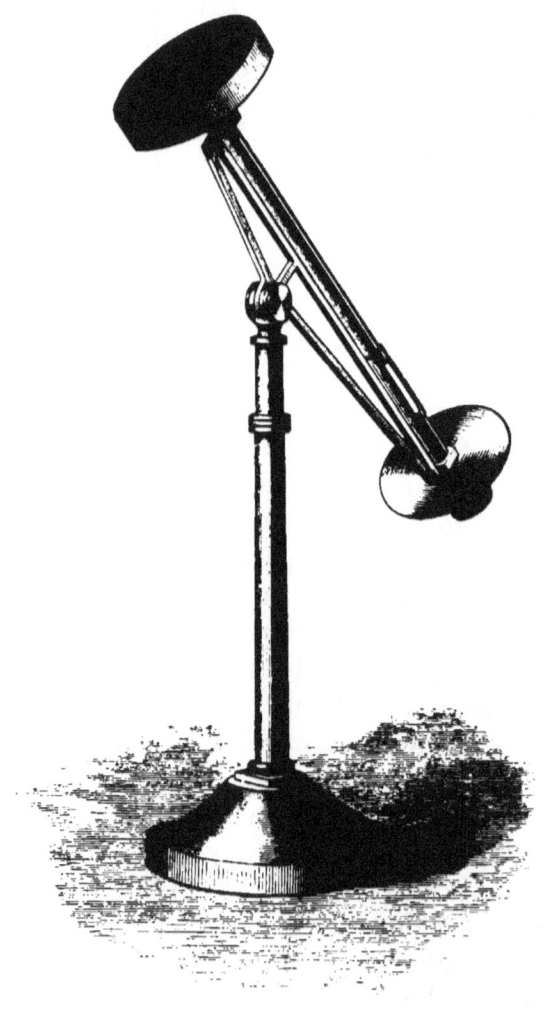

cylindrical part, and on the back, because this is an exceedingly bad radiator of heat, so that the instrument loses by those sides very little of the heat which it collects by the blackened side, which is a good absorber and is turned directly to the sun. This little silver vessel is filled with water, and all the radiant heat and light, everything in the form of radiation that falls upon this lampblack, is absorbed by it, and is degraded into the form of heat and so communicated to the water. In the middle of the water is the bulb of the thermometer, whose stem runs down through the axis of the apparatus. We can adjust it so that the blackened disc shall receive the sun's rays perpendicularly, by a very simple contrivance :—a disc of metal at the other end of the thermometer tube, of exactly the same size as the first disc : then the whole being so set that the shadows of the two discs coincide,—we know that it is turned directly to the sun. Take off the cap of the instrument for a measured period, put it on again, and after the whole has been thoroughly shaken up, so that the temperature of the water is the same throughout, read off the rise of temperature as shown by the thermometer. Correct that for the loss of heat by radiation during the performance of the experiment. That can be done at once by simply watching how it gradually loses heat when it is turned to the sky, but screened from the sun's radiation. With this instrument we can make a fairly approximate estimate of the amount of heat which is received from the sun by the blackened surface in a given time ; and by comparing the surface of this disc with the surface of the whole earth which is exposed to the sun, we can estimate at least approximately how much radiant energy in the form of heat, light, actinism, and so on, comes to us per second from the sun ; and therefore we can estimate what amount of energy leaves the sun's whole surface every second,—that is to say, what number of foot-pounds of energy the sun is spending per unit of time.

According to Thomson (calculating from the data of Pouillet and Herschel), the sun's radiation is equivalent to about 7000 horse-power per square foot of his surface—somewhere about thirty-fold that of the same area of the furnace of a locomotive—and somewhere about 6×10^{30} units of heat (C.) leave his whole surface per annum.

In addition to the data which I have just given you, I shall conclude this morning by giving one or two others. Let us take the case of the earth's motion in its orbit. The immense mass of the earth moving round in its circle of over 90,000,000 miles radius in one year is moving at what we should consider an enormous rate, far greater than that of a cannon ball (being in fact about 80 times as great), and yet the whole kinetic energy it would supply, if it were accidentally to impinge upon a huge target,—as an Armstrong projectile goes against an iron plate,—is a mere trifle to what we have been considering; it could only supply by that frightful crash an amount of heat equal to the sun's loss in about 80 days. But if instead of taking its energy of motion in its orbit, you were to take its potential energy, as a heavy body which, if allowed, would fall into the sun, and there produce an immense development of heat by impact, the calculation leads us to this result, that it would acquire, on reaching the sun's surface, such a speed that the energy of the impact would be equivalent to the heat at present given out by the sun in about 91 years. But the planet Jupiter is not only enormously more massive than the earth, but is also very much farther away from the sun, and therefore on both accounts it would produce a much greater development of heat if it were to fall into the sun. The calculations made on the same data for the planet Jupiter give something like 32,000 years,—that is to say, Jupiter alone falling into the sun would supply its present loss for 32,000 years to come.

Then, there is one final datum with which I shall conclude to-day, and it is this:—I shall give more detailed explanation of it in my next lecture, but I wish to mention it before concluding,—that the lowest possible estimate which we can make of the capacity of the sun for heat is such that, cooling at the present rate—losing energy at its present rate—the sun cannot possibly cool more than a single degree Centigrade in seven years. It may be, on the highest estimate we can take, one degree in 7000 years; the data are very uncertain; but we may say that these are the limits between which it must lie. Startling as are many of the matter-of-fact statements I have made to you to-day, I cannot help once more repeating this, by far the strangest of them all: the sun has such an enormous capacity for heat that it takes at least seven years, at its present enormous rate of radiation, to cool by one degree Centigrade!

LECTURE VII.

SOURCES AND TRANSFERENCE OF ENERGY.

Available Sources of Energy on the Earth. Whence these have been derived Uniformitarian School of Geologists. Sir W. Thomson's arguments as to the length of time during which life has been possible on the earth. Transference of Energy—through Solids, Fluids, and through the Ether. Test of the Receptivity of a body or system for energy in a vibratory form Physical Analogies introductory to Spectrum Analysis.

IN my last lecture I considered, in as great detail as our necessarily limited time permitted, the origin of the energy of the solar system. I must now consider in part of to-day's lecture a smaller, but much more important matter,—much more directly personal to us,—namely, our available sources of terrestrial energy. In my little work upon Thermo-Dynamics, I have arranged these sources in order as follows :—

First. Our available sources of potential energy.

1st, Fuel. Under the head of fuel I should include not merely coal, wood, and so on, but also all that may properly be called fuel—the zinc used in a galvanic battery, for instance, and various other things of that kind.

2d, The food of animals.

3d, Ordinary water-power.

4th, Tidal water-power.

All these are forms of potential energy.

Then, *Secondly*, in the Kinetic form, we have—

(1.) Winds.

(2.) Currents of water, especially ocean currents; and finally we have

(3.) Hot springs and volcanoes.

There are other very small sources known to us, exceedingly small; but these I have named include our principal resources.

Now comes the question, what are the sources of these supplies themselves? I find I have classified them also under four heads.

The first is primitive chemical affinity,—chemical affinity which we may suppose to have existed between particles of matter from the earliest times, and still to exist between them, because these portions of matter have not combined with one another nor with other matter. If, for instance, when the materials of which the earth is at present composed were widely separated from one another, there were particles of meteoric iron and native sulphur which, when the materials did come together to form the earth and heated one another by mutual impact, did not combine together but have still remained through long periods of time separate from one another, we should consider that the mutual chemical potential energy of the iron and sulphur remains to us as a portion of energy primordially connected with the universe. But of that, so far as we know, at least near the surface of the earth there is very little. There may be towards the interior enormous masses of as yet uncombined iron and uncombined sulphur, or various other materials, but towards the surface, where they could be of any direct use to us, the quantities of these are excessively small.

The second source is that which I have several times alluded to,—solar radiation,—and that is by far the most abundant source we have.

Then we have two very instructive forms, viz., the energy of the earth's rotation about its axis, and the internal heat of the earth.

Now, if we take in turn the enumeration which I gave at first of our available stock, and compare it with the sources from which we derive that stock, we shall easily see how the two are connected with one another.

First, we have fuel. Now, our supplies of fuel are almost entirely due to the sun. That is to say, in times long gone by, the sun's rays by their energy, as absorbed in the green leaves of plants, decomposed carbonic acid and stored up the carbon. That carbon, and various other things stored up ages ago along with it, we have still as an immense reserve fund of coal.

Then for the food of animals we are mainly indebted to the sun again, because the food of animals must ultimately be vegetable food, even of the animals which live upon animal food. Then for ordinary water-power we are also indebted to the sun, because it is mainly the energy of the radiation from the sun in its heat form which evaporates water from the plains or seas, and allows it to be precipitated again at such a height that it has potential energy in virtue of its elevation. Ordinary currents of water are a mere transformation of this potential energy, because water on a height may convert part of its potential energy into kinetic energy of visible motion as it flows down.

But when we come to tidal water-power we must look to another source. If we employ tidal power for the purpose of driving an engine, we take it in the rise of the water as the tide-wave passes us. We secure a portion of water at a certain elevation, wait till the tide has gone back, and then take advantage of the descent of that portion of water. Now, if we were to go on doing that for any considerable period of time, and doing it over large tracts of sea-coast, we should find that the effect of it in time would be to gradually slacken the rate of rotation of the earth; so that if all our important sources of power, such as coal, and direct solar radiation, were to fail us in great part, and if we were driven finally as a last resource to use tidal water-power, it might come to be a very serious international question between those kingdoms which possessed sea-board and those which had none. For if it were largely employed, the period of the rotation of the earth might be in a moderate period of years seriously affected. And there seems to be no known compensating advantage for those nations who are not possessed of an extensive sea-board within the Temperate or Torrid Zones, where alone this source of power would be of much avail.

Then we have, next to these, winds and ocean currents. These are almost entirely due to solar radiant heat. And, finally, hot springs and volcanoes, which have never been employed for any direct production of work, but which might possibly be so used. Their energy depends, mainly at least, upon the internal heat of the earth; partly perhaps on potential energy of chemical affinity.

So you see that mainly to solar radiation, but partly to the other three sources of supply, are due the various stores of energy which we have at our disposal. This, however, is a mere bare enumeration. I might spend many lectures developing small parts of this grand subject; but I have given you in these few words the large heads, and it is scarcely compatible with the time at our disposal to devote another couple of lectures to pursuing the subject into its minute details.

The next question I take up is this,—intimately connected with what we have just considered: the question of how long something like the present state of things has been going on on the earth's surface. This is an extremely important question, and can be approached from various sides,—from the geological side, for instance, by consideration of the thickness of strata, of amounts of erosion, and such like; but it can also be approached directly from the point of view of energy, and from that point of view alone I shall now attempt briefly to treat it.

The old notion of what was called the Uniformitarian school of Geology, was simply that things had been going on and were to go on, both in the past for many millions of years, and in the future for many other millions of years, at as nearly as possible the same uniform rate,—that we were getting a steady supply of heat from the sun,—that even if energy (it was not called energy in those days), even if some source of supply, call it what you like, was disappearing in some portion of the interior of the earth, at its disappearance it was producing say electric currents, and decomposing some compound substance; so that, if ever lost by chemical combination at one place, electric currents would be produced, and something equivalent thereby given out in some other place, so that the stock should be maintained as nearly as possible at a uniform state.

Now, this is totally inconsistent with modern physical knowledge as to the dissipation of energy. Transformations must be going on now (at least on the average) at a much slower rate than they were going ages ago. Just as when you take a red-hot ball from a furnace; it cools at a certain rate, but as it becomes colder it cools more and more slowly. And this is not a mere analogy, but an almost absolute identity, with the case of the earth and the sun. There is no doubt that at some period long ago the earth was so hot as to be at all events plastic, if not absolutely liquid throughout its mass; and there is no doubt that at the present moment, even after ages of expenditure of energy at a very great rate, the sun must be still liquid in great part, and even gaseous in very large part.

Now, we can apply the theory of energy, especially from Carnot's point of view, to the state of things in the earth and in the sun, and can at all events roughly approximate to the period during which the earth has been habitable for animals and plants such as we find upon it now. We do not say, of course, that it *was* inhabited for such periods by animals and plants such as we see now, or find fossil remains of; but we can trace approximately backwards for how long the earth was habitable by such, and that is the problem we propose to solve.

This subject was taken up very carefully within the last few years by Sir William Thomson, and the brief *résumé* I shall give of his results contains nearly all that is accurately and definitely acquired to science upon the subject. He divides his arguments upon it into three heads. The first is an argument from the internal heat of the earth; the second is from the tidal retardation of the earth's rotation; and the third is from the sun's temperature.

Now, as regards the internal heat of the earth, we know by actual observation that as we go down a deep mine we find the temperature almost invariably increasing. We know also that whenever a body is hotter at one part than another, the tendency of heat is always to flow from the hotter part of the body to the colder. Therefore, as the earth's crust is warmer and warmer as we go farther and farther down, there must be a steady flow of heat outwards from the interior to the surface. The earth is therefore even now losing heat at a certain perfectly measurable and calculable rate. But if it is losing heat now we can calculate by known physical laws and known mathematical processes, from the present state of distribution of temperature,—we can calculate backwards how its heat was arranged a hundred thousand or a thousand thousand years ago, just as certainly—if physical laws as we know them now were in existence in the past—as we can predict from our mathematical calculations what will be its distribution at any time future, if these physical laws continue to hold. In working out such a question as this, it is found that the rise of temperature, taken (over the whole earth's surface) at an average of about one degree for 100 feet of descent, leads to this conclusion, that about ten millions of years ago the surface of the earth had just consolidated, or was just about to consolidate; and in the course of a comparatively few thousands of years after that, the surface which had been consolidated had become so moderately warm as to be fitted, at all events in some parts, for the existence of life such as we know it. That is to say, the surface temperature, in certain regions at least, was not greater than that which is perfectly easily borne by animals and vegetables in the tropics at the present day; and the rate of increase of temperature in going down below the surface was one degree in perhaps every six inches, or every ten inches, or something of that sort. That would not interfere very greatly with the growth of vegetables; so from this point of view we are led to a limit of something like ten million years as the utmost we can give to geologists for their speculations as to the history even of the lowest orders of fossils.

If we were to trace the state of affairs back, instead of to ten millions, to a hundred millions of years, we should find that (if the earth then existed at all, if that collocation of matter which we call the earth was then actually formed), and if the physical laws which at present hold have been in operation during that hundred million years, then the surface of the earth would undoubtedly have been liquid and at a high white heat, so that it would have been utterly incompatible with the existence of life of any kind such as we can conceive from what we are acquainted with. Thus we can say at once to geologists, that granting this premiss,—that physical laws have remained as they are now, and that we know of all the physical laws which have been operating during that time,—we cannot give more scope for their speculations than about ten or (say at most) fifteen millions of years.

But I daresay many of you are acquainted with the speculations of Lyell and others, especially of Darwin,[1] who tell us that even for a comparatively brief portion of recent geological history three hundred millions of years will not suffice!

[1] *Origin of Species* (1859), p. 287.

We say—So much the worse for geology as at present understood by its chief authorities, for, as you will presently see, physical considerations from various independent points of view render it utterly impossible that more than ten or fifteen millions of years can be granted.

You see, then, that the argument from the internal heat of the earth depends upon working the problem backwards, and finding what is the utmost limit of time back at which the surface of the earth could possibly have been fitted for the life of either animals or plants.

And this leads me to say a word or two about one of the most remarkable results of investigations of this kind,—investigations conducted as purely mathematical problems, and based entirely upon physical experimental data, viz.,—upon the observed laws of conduction of heat. In the great majority of problems where the data are of the nature of those we have as to the underground temperature of the earth, the question of the future is a perfectly definite one. If we knew the present thermal condition of every part of the earth's mass, we could calculate what would be the temperature at any depth below the earth's surface at any time future, provided things went on under the same conditions as they are going at present, and our results would be always perfectly and directly intelligible. But when we try to work the problem the other way, when we ask what must have been the thermal state of such a body as the earth at such and such a time past, then we invariably, or almost invariably, find a limit of time beyond which our equations become uninterpretable. So far as our equations represent what would be the course of nature provided the existing physical laws remained true, there must have been at this definite epoch of past time the introduction of a new state of affairs, something which arose from a previous state by means of a process not contemplated in our investigation.

In the case of the earth there is no particular difficulty in understanding what might have been that anterior state of affairs. We can trace matters back to the time when the earth was molten throughout. Going farther and farther back, we come to a distribution (which might be pretty nearly uniform) of heat throughout the whole mass. Now, a uniform distribution of heat throughout the whole mass could have had no existence for more than an instant, so far as we know; and we cannot conceive it to have arisen from any previous distribution of heat in the mass. But we can understand how a high temperature throughout the whole mass might have been produced by the materials of which the earth is composed falling together. If they fell together in such a way that the whole mass of the earth was agglomerated together almost at once; and if the different parts impinged together with properly arranged velocities, it is possible the earth may have been agglomerated together, so as to have for a moment the same temperature throughout, thus giving us something like what we have deduced from our formulæ. But you will notice the state of things before and after that moment. Before that moment it was cold masses of matter, separated perhaps by millions of miles, or far more than that, but having potential energy of gravitation gradually being transformed into kinetic energy of approach. Then, at the instant of impact, came the critical change. Instead of the cold scattered masses of matter, there was suddenly an agglomerated mass of almost uniform temperature throughout, and it has been cooling and shrinking ever since.

The second of these arguments of Sir William Thomson depends upon the tidal retardation. In my first lecture I mentioned to you that there was such an effect, and that it had been actually observed by astronomers in a very peculiar way; because, on calculating back from the known present motion of the moon, it was found that there must be some unrecognised peculiarity in that motion, which had not been deduced by calculations founded upon gravitation, either as attraction or as disturbance. The moon, in fact, seems to have been moving quicker as time has gone on, since the eclipses of the fifth and eighth centuries before our era. The only way, as Laplace put it, in which it could be accounted for in his time, was by what he called 'secular acceleration of the moon's mean motion.' In other words, the average angular velocity with which the moon moves round the earth appears to have been increasing for the last 2000 years or more. He showed that there was a mode of accounting for this by planetary disturbance of the earth's orbit, and as calculated by him, this explanation seemed to account for exactly the amount of acceleration which was observed in the moon's motion. Using his formulæ and the numbers calculated from them, and working back to those old days, we find we arrive at almost the circumstances of those eclipses as described by historians.

Fortunately, Adams, a few years ago, revised Laplace's investigation, and found that he had neglected a portion of the necessary terms, and that the explanation given by Laplace, when properly corrected, accounted for only one-half of the phenomenon observed;

so that there still remained one-half of the quantity to be accounted for. This could not be accounted for by the disturbance of other bodies attracting the moon. Why then does the moon appear every revolution to be moving faster and faster round the earth? Well, the only way in which we can explain it, after we have made every possible allowance for effects of disturbance by other planets, is simply to inquire, Does our measure of time continue the same?

We measure the time of the moon's revolution in terms of hours, minutes, and seconds; but these hours, minutes, and seconds are measured for us not by our clocks, as you may at first think. We set our clocks by the earth's rotation, and therefore it is in terms of the earth's rotation that we measure the time of the moon's revolution round the earth. So that the moon will appear to be moving quicker round the earth, even supposing her orbit be altogether undisturbed, if the earth itself, which is furnishing the unit of time in which her revolution is to be measured, is rotating slower and slower from age to age.

Then comes the question, Is there a cause which tends to slacken the earth's rotation? Newton laid it down, in his First Law of Motion, that motion unresisted remains uniform for ever, and referred to the earth as a particular instance where there is nothing in the attraction of the sun or moon, or the disturbance caused by any of the other planets, affecting the rate of its rotation about its axis. But it was left to Kant, first of all, to point out, and even to approximate in amount to, a resistance to the earth's rotation due to the tide-wave; and to show that the earth, because the tide-wave is lifted up towards the moon, and on the opposite side from the moon, has constantly to rotate inside what is practically a friction-brake. The water is held back by the attraction of the sun and moon, and the earth has to move inside this shell of water. There is therefore a source of constant friction, and friction of course constantly produces development of heat. The heat must be accounted for by some energy transformed, and what is here transformed is part of the energy of the earth's rotation about its axis. So long as tides go on, there will therefore be constantly a retardation of the rate of the earth's rotation.

Now let us see when this relaxation of the earth's rotation would cease. Obviously this would be at the instant when the earth at last ceased to rotate within the tide-wave; in other words, when the tide-wave rotates along with the earth, when it is always full tide at one and the same portion of the earth's surface, the tide-wave being fixed (as it were) upon the earth's surface. But the tide-wave is always, approximately at least, directed towards the moon, so this part of the surface where the tide-wave is fixed for ever must be constantly turned towards the moon. In other words, if there were no sun producing tides, but the moon only, the final effect of the tides in stopping or quenching the earth's rotation would be to bring the earth constantly to turn the same portion of its surface towards the moon, and therefore to rotate about its axis in the same period as that in which the moon revolves about it. This most remarkable ultimate effect we see already produced in the moon,—it is precisely the same thing,— we see the moon turning almost exactly the same portion of its surface to the earth at all times. The little deviation we see occasionally is precisely accounted for by the fact that the moon's orbit is not exactly a circle, and therefore the moon does not move in it with the same rapidity when it is nearest the earth as it does when it is farthest away from the earth. We are thus, as it were, enabled occasionally to see a little round the corner. The moon is now rotating precisely in the way in which the earth will in time rotate when as much as possible of its energy of rotation is used up in producing heat by tidal friction. And that the moon should already have come into this state so long before the earth has arrived at it, need not surprise us. The moon's seas (when she had them) were of molten lava,—far more viscous than water; the tide-raising force on her surface depended on the mass of the earth, some *eighty* times greater than that of the moon, which is the main agent in our comparatively puny tides: and, in addition, the moon's moment of inertia is very small compared with that of the earth.

It being thus established that the rate of rotation of the earth is constantly becoming slower, the question comes: How long ago must it have solidified in order that it might then have the particular amount of polar flattening which it shows at present? Suppose for instance it had not consolidated less than a thousand million years ago. Calculation shows us that at that time, on the most moderate computation, it must have been rotating at least twice as fast as it is now rotating. That is to say, the day must have been 12 hours long instead of 24. Now, if that had been the case, and the earth still fluid throughout, or even pasty, that double rate of rotation would have produced four times as great centrifugal force at the equator as at present, and the flattening of the earth at the poles and the bulging at the equator would both have been much greater than we find them to be.

We say then, that because the earth is so little flattened it must have been rotating at very nearly the same rate as it is now rotating, when it became solid. Therefore, as its rate of rotation is undoubtedly becoming slower and slower, it cannot have been many millions of years back when it became solid, else it would have solidified into something very much flatter than we find it. That argument, taken along with the first one, probably reduces the possible period which can be allowed to geologists to something less than ten millions of years.

Then comes the third argument,—it is not quite so emphatic in its demands for restricted periods as either of the other two,—the argument from the length of time that the sun can be imagined by its radiation to have kept the earth in a state fit for the habitation of animals and vegetables. The argument from this point of view, I say, is not so trenchant as the others, because we can imagine that when the sun was immensely hot, as it must have been at some previous time,—enormously hotter than at present,—we can imagine that one effect of its heat was to throw off from its surface such enormous clouds of absorbing vapour, which cooled as they left the surface, that the effective amount of radiation reaching the earth might not have been greater than at present. So it is possible to conceive a sort of uniformitarian state of radiation from the sun:—accounting for it by saying that when the sun was hottest and was radiating the most, it was simultaneously raising the greatest amount of obstructions to the propagation of radiations from its surface. A similar argument might, of course, be devised with reference to the greater amount of vapour which increased solar radiation would raise to be condensed in the earth's atmosphere. However, if we make the supposition that the sun has been cooling even at a uniform rate, we find that this mode of calculation leads us, in spite of the enormous amount of heat which must have been produced in the sun by the impact of its materials when they fell together, to the conclusion that on the very highest computation which can be permitted, it cannot have supplied the earth, even at the present rate, for more than about fifteen or twenty million years.[1]

This, I again say, is not so trenchant an argument as either of the other two; but the conclusion from these three arguments is not, as some of Thomson's opponents seem to imagine, only as strong as the weakest of the three. In order to upset the conclusions drawn from them, it would be necessary to disprove two of these arguments, and greatly to damage the third. But each of these arguments is quite independent of the other two, and is—for all tend to something about the same —to the effect that ten millions of years is about the utmost that can be allowed, from the physical point of view, for all the changes that have taken place on the earth's surface since vegetable life of the lowest known form was capable of existing there.

I leave this part of the subject for a time. This has been a developed application of the theory of energy

---

[[1] *Note to Third Edition.* Several critics, as well as some writers of a higher order, think they have detected inconsistency between this passage and another in p. 156. There is no such inconsistency. At p. 156 the *whole* supply was spoken of; while here we are dealing with what has been *already* expended.]

to the solar system first, and then in particular to our own earth.

Now, I pass to one or two other applications of the second law of thermodynamics, especially in the beautiful part of it furnished by Carnot's reasoning. We have now to take up the consideration of the transference of energy from one body to another, not the passage of energy from one part of a body to another portion of the same body. That is in the main the question of the conduction of heat, to which I shall devote another lecture. But now we are to speak of the radiation of heat and light from one body to another.

But before I take up that I shall direct your attention to one or two experiments, some of them long known but at their epoch hardly explained, others only recently made.

First of all, let us take as the medium of communication between two bodies:—the medium through which the energy is to be transferred from one body to another:—a strong wooden framework such as this. I have two pendulums with very massive bobs suspended from it, and have carefully made these two pendulums as nearly as possible of the same length, so that their times of vibration are as nearly as possible the same. Both pendulums are now at rest, but suppose I set one to vibrate, leaving the other at rest, you will notice, if you watch the second for a short time, that it begins to vibrate in its turn, and as time goes on it swings through larger and larger arcs of vibration, till at last the first pendulum is reduced to rest. Now, this is quite obviously a case of transference of energy from one pendulum to the other, effected, you will see, through the wooden structure; but it has been effected thus completely on account of the simple fact that the two pendulums had been (as it were) previously tuned together and made to vibrate in precisely equal times. We shall presently try the experiment with the two pendulums not tuned together, and then you will see

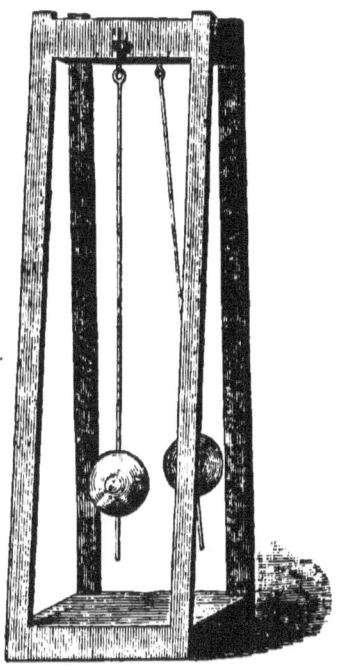

that there may be transference of energy for a few minutes, but it will be far less complete, and in the course of a very short time the whole will be given back again to the first pendulum, and so on. In the case before us, a short time will suffice for the whole of the energy to be transferred from the one pendulum to the other, and it will then be just as if we had turned the whole apparatus round through two right angles. You will have the second pendulum vibrating with the whole original energy in place of the first, then the transference will go on again in the opposite direction, and the first will get back what it lost, except what has been unavoidably dissipated in producing air vibrations, and in producing heat in the materials of the framework, which is not a perfectly elastic body, and all throughout which friction and various other disturbing causes operate. Notice particularly that the mode of transference in this case is through a solid body, and that it is simply by vibration of the solid body that it has been effected.

I pass from the consideration of transference through a solid body to transference by a gaseous body; and we shall easily realise precisely the same effect by means of a couple of tuning-forks. These forks are tuned precisely to the same note. They are furnished with resonating cavities, to enable them to communicate to the air as much of their energy as possible. If I set one in vibration, the effect of the resonating cavity is to enable it to set in lively motion, at its own period of vibration, the air surrounding it. But here is another cavity which is tuned to that particular time of vibration. The tuning-fork attached to it is also tuned to precisely the same note, and now we find that when I first of all start the first tuning-fork, then turn it so as to place its resonating cavity with the mouth towards the mouth of the resonating cavity of the other, through the gas-filled space between the two, there is a transference of energy which is such that if, after a second or two, I suddenly stop with my finger and thumb the vibrations of the first fork once for all, you will hear the other resounding with considerable loudness. The transference of energy has here been made through the air instead of through a solid body, as in the case of the wooden framework connecting the pendulums.

[I now call your attention once more to the massive pendulums, because the first has again handed over the greater portion of the energy to the second. My assistant will now put them out of tune, and we will try the experiment again.]

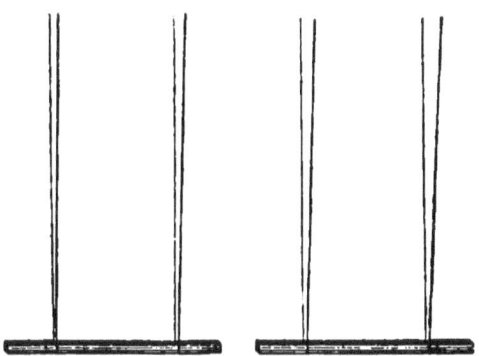

Connected with these, and to be explained on precisely the same physical principles, we have another strikingly illustrative experiment. Consider this third arrangement, where we are to have the transference of energy effected, not as in the case of the pendulums through a solid bar, nor as in the case of the tuning-forks, through the gaseous medium between the two, but simply by magnetic action :—force acting between a couple of steel bars; an action which, as you all know, is not affected by the interposition of any non-magnetisable body whatever, and which is as energetic through what we call a vacuum as through air. These bar magnets are as nearly as possible of equal mass, and are supported by strings or wires of equal length. If I take one of them away, the time of oscillation of the other will be the same, whichever I take. In their position of equilibrium they hang in the same horizontal line. Now they are both at rest at this moment. Suppose I communicate vibration to one of them in the direction of its length. You notice how very rapidly the energy is transferred from the one to the other. The magnet which was at first at rest has now gained the greater part of the energy, and in the course of a very few seconds more you will see the other has lost it all. There it is: absolutely at rest for a moment; and now the process recommences the other way. After exactly the same interval of time as that which elapsed from the commencement of the experiment to the instant of the first magnet's being brought to rest, the second will be brought to rest in its turn. There it is at rest now —for an instant only; and the same transference will go on again indefinitely. Now, what is it that conveys the energy in this case? The transference of energy is due entirely to the magnetic attraction of one of those bars for the other; because, though the apparatus is constructed suspiciously like that which I employed a few minutes ago for the massive pendulums, the masses of these bars are not sufficient to produce any appretiable effect upon the supporting beam, so that it would be impossible, if we were to demagnetise these bars, to obtain any appretiable transference of energy from the one to the other. This then is transference of energy from one body to another, not through a solid, nor through a gas, as in our recent experiments, but through the magnetic medium, whatever that may be,—what Clerk-Maxwell has given us strong reason to believe is the same medium as that which conveys light and radiant heat. So we have here, as it were, a third mode of transference of energy from one body to another; and this resembles much more nearly than either of the other two the cases to which I am about to proceed.

[But before I so proceed, you will notice that I have got the original pair of massive pendulums on the wooden frame put out of tune, and you can now study how the oscillations are handed on from the one to the other. You see that the transference, if any at all, is very much more slight than before, and not only is it slight, but after a short time it ceases, and then sets in the other way. The energy of the second pendulum is sometimes falling and sometimes increasing, but it never rises to any great percentage of what remains in the first. In fact, because of the dissimilarity of their periods of oscillation, the one comes sometimes into a position in which it can gain energy from the other, and a second or two later it puts itself into such a position as to lose energy, and so on backwards and forwards; whereas, when the two were tuned almost exactly to one another, if they were at any instant in such a position that the one was giving energy to the other, they would remain for a very long period in such a relative position. The one would always be throughout that period in the most favourable position for communicating energy to the other, and this solely because their periods of oscillation were alike; whereas when their periods of oscillation differ, the one is sometimes getting away from the other, and sometimes getting pulled back.]

All of you must have noticed this in the ringing of a massive bell. Even a child can ring an immensely massive bell with very slight application of force, provided he perseveres in pulling exactly at the proper moments. Just as the bell is about to descend, let him pull, so as to quicken the motion, but let him slacken when the motion is such that a pull would tend to stop it. By waiting till the exact moment, and properly timing the impulse, he is capable of giving large oscillations to a mass which otherwise he is almost incapable of setting in motion.

In the same way it is possible to check it by applying retardations exactly at the proper moment. This would be at exactly equal intervals of time, representing the vibration of the bell if it were left to itself.

Thus all these experiments depend upon the transference of energy in a kinetic form between two bodies, and the test of the capability of the one for receiving the energy which is sent out by the other is this, that the natural undisturbed times of vibration of the two bodies shall be as nearly as possible precisely the same. I have not time to enter more deeply into the subject to-day, but I shall endeavour, in the few minutes which remain to me, to sketch briefly what is to be our application, to modern science, of these purely mechanical experiments.

Suppose we have a substance which, instead of giving off sound, in consequence of its vibrations, is vibrating so rapidly as to be giving off some particular colour of light or of radiant heat. Then the substance which will be best qualified to absorb that particular colour of light or of radiant heat, will be another body of precisely the same kind as the first, because the two specimens of the same matter will, under the same circumstances, vibrate according to precisely the same laws ; and therefore if you define a particular beam of light by having it sent out from a particular substance which is rendered incandescent, another specimen of the same substance will find in the beam precisely those particular times of vibration which most aptly suit it, and therefore will be best fitted to absorb them.

This is, briefly, the dynamical principle at which Professor Stokes arrived more than twenty years ago, and which, if its applications had been properly pursued at the time, would have given us ten years' start in our knowledge of celestial chemistry. Stokes' illustration was this :—He imagined a space, such as this room for instance, to be filled with tuning-forks (with resonating cavities let us say) or with pianoforte wires stretched about in all directions so as not to interfere with one another, but as nearly as possible to fill the whole space. If all the tuning-forks, or all the pianoforte wires, are tuned to the same note, that arrangement will form a medium which is capable, when agitated in the simplest manner, of giving out only that particular note. Set all the tuning-forks to vibrate, they all conspire to strengthen one another and give out their one particular note, and that note only. On the other hand, when you use that arrangement not as a source of sound, but as a medium through which you endeavour to make sound pass, then from what I have just shown you, you will obviously find it to be particularly opaque to that particular note, and to that note only. Suppose a performer with a powerful instrument (such as a cornopean) placed at one side of the room, and a listener at the other. Then let the player play any note he pleases except the note belonging to the forks or strings, that note will be heard in full intensity, except in so far as the strings (merely as obstacles) intercept the passage of the sound. Such a note will be heard almost as powerfully on the other side of the room as if there had been no tuning-forks or wires present. But as soon as he plays the particular note which belongs to all the forks or all the strings, it comes to be just the question of the pendulums or magnets, or the two tuning-forks which I have just shown you. The contents of the room gradually absorb each a portion of the sound which reaches it, and are set into vibration by it. If there be enough of them they take all the energy of the sound, and of course completely prevent the sound from passing through the medium, except in so far as they give it out themselves.

Here, then, is a medium which of itself can give out one definite note, and one note alone, when it is a source of sound; but which, when it is employed as a sort of sifter of sound, can sift out from a mixed or confused sound only that particular note. That then is mechanically or physically the analogy to which we shall have to reduce the fundamental principles of spectrum analysis.

LECTURE VIII.

RADIATION AND ABSORPTION.

History of the discovery of the Physical Basis of Spectrum Analysis. First result of Spectrum Analysis applied to non-terrestrial bodies;—There is Sodium gas in the Sun's Atmosphere. Elaborate experiments of Stewart and Kirchhoff. Identity of Light and Radiant Heat. Distinctive characters of a particular ray. Application of Carnot's principle to establish the equality of radiating and absorbing powers. Black, transparent, and perfectly reflecting bodies.

I ENDED my last lecture by considering various modes of transference of energy of vibration from one body to another. I took in particular three cases, in the first of which the transference took place through a solid body, in the second the vehicle was ordinary air, and in the third case it was the medium which propagates magnetic and electric actions. But in every one of these cases we found that the condition which is absolutely necessary for a complete handing over of the energy of one vibrating body to another, whatever be the intervening medium of communication, was that the time of vibration of the second body should be adjusted to be exactly equal to the time of vibration of the body which had the energy at first.

I then went on to suppose a finite space to be filled with a number of such vibrating bodies, all tuned (as it were) to vibrate in precisely the same time; and I showed you that if we considered a space so filled to act as a medium, it would be such as when set in vibration would give one perfectly definite sound or note of one definite pitch, and that it would be competent on the other hand to absorb precisely that particular note and no other. All other musical sounds would pass through it without marked interruption unless its transverse dimensions were very great; but if that particular note were played in its neighbourhood none of its vibrations could pass through the medium. They would, if the medium were only deep enough, be almost entirely absorbed and retained by the medium.

Now, that is the dynamical principle which led Professor Stokes, about the year 1852, to the first distinct anticipatory statement of the physical basis of spectrum analysis. In order that I may make quite intelligible to you how he made the application to the analogy between the behaviour of certain bodies as regards light and the behaviour of those tuning-forks and strings as regards sound, it will be necessary for me to make a slight digression. That digression has reference to the different refrangibilities of the different colours or wave-lengths of light.

It was one of Newton's simplest and yet greatest discoveries in optics that when a beam of white sunlight passes through a prism it is divided into its various components. The components have different wave-lengths, or different times of vibration, or different refrangibilities,—for these three properties vary together,—in virtue of which their paths, when they pass out of the glass prism, differ in direction; and therefore they all spread out from each other into the graduated series of colours which Newton called the solar spectrum. Newton's first method of forming the spectrum was complete so far as the object he had in view was com-

carried; but it is not sufficient to enable us to make that searching scrutiny of the composition of sunlight which alone would enable us to tell whether certain perfectly definite coloured rays are wanting in it or not. In order to achieve this it is absolutely necessary to make some optical arrangement by which no two portions of coloured rays of different refrangibilities shall be allowed to overlap one another, as it were, in the spectrum.

Newton's first method was simply to cut a round hole in the window-shutter of a dark room, and allow the beam of sunlight which entered by it to pass through his prism, and then to be received upon a white screen. Make then, the simplest possible supposition, so as to save complexity at the commencement of our explanation:—suppose that there were only two distinct kinds of homogeneous light in sunlight,—that it were made up, for instance, of homogeneous red and green in proper proportions to make white. By this method of experimenting we should have had on the screen, before inserting the prism, a circular spot of white sunlight. After the interposition of the prism this would be decomposed into two circular spots: a red spot depending on the one kind of light, and a green spot depending on the other; both displaced, but the green most, from the original position of the white spot. But, as we know, sunlight consists not merely of a particular kind of red and a particular kind of green, but of almost every shade of colour intermediate between and beyond these limits on each side; and it is therefore evident that the method gives superposition of equal round spots of light of gradually increasing refrangibility, with their centres arranged continuously (or almost continuously) along a straight line on the screen; so that there must be a constant overlapping of a great many of these successive spots at any one point of the spectrum, and therefore it must be practically impossible by this method to detect the absence of any one particular shade of colour.

Now, though the optical method which Newton[1] devised for the purpose of avoiding this difficulty is a very simple one, it deserves a word or two, as it will help you to understand the experimental illustrations I mean to give in my next lecture. Instead of using a round hole we now use a narrow slit whose sides are perfectly parallel to each other, and which can be made (by proper mechanical adjustment) as wide or as narrow as we choose. The light from the sun or electric lamp, or whatever source we employ, comes through this slit as a thin sheet, and falls upon an achromatic lens; that is, a lens which behaves in almost precisely the same way to all the differently coloured rays falling upon it. It is usually convenient to place the lens at such a distance from the slit that at exactly the same distance beyond the lens an image of the slit, equal to it in size, will be formed on the screen. If, then, sunlight pass through the slit, and fall upon the lens, we shall have, on a screen placed at the proper distance, simply an intensely bright white line, consisting of all the different rays belonging to sunlight. But if you interpose in the course of those rays, just after they come through the lens, a prism, with its edge parallel to the slit, the effect will be a change of direction of those cones of rays which are converging towards the image. The prism will most refract the

[1] *Optics*, Book I. Part i. Exp. II, Illustration.

violet rays, and all the others will be less and less refracted as their wave-lengths grow longer and longer till we reach the lowest red in the spectrum; and therefore instead of having a set of coloured discs, as by the first method, succeeding one another with their centres along a line, and overlapping, you will have a set of parallel coloured images, each no broader than the slit itself, and you can make the slit as narrow as you please. In every part the consecutive images lie side by side, contiguous to one another; but if there be light of any wave-length or any particular refrangibility which is wanting, then the space corresponding to that will be left as a dark line (an unilluminated image of the slit) across the otherwise continuous coloured band.

You see hanging on the wall a coloured plate representing the solar spectrum, formed in the way I have just pointed out, and you can see those dark lines across it. Only a few of the chief ones are figured. The number of those whose position is already carefully measured, or photographically registered, amounts to many thousands. [See diagram, p. 192.] They were first noticed by Dr. Wollaston, about the beginning of this century, but he paid very little attention to them; and they were re-discovered a considerable time afterwards by the great optician Fraunhofer, whence they have been called Fraunhofer's lines.

Of those lines one of the most remarkable is that to which Fraunhofer gave the name of D, which you will see upon the picture near the boundary between orange and yellow. When, however, a very perfect prism is used, and a telescope is employed instead of the screen to receive the spectrum, then we are enabled to see that this line is double. This gives it a very remarkable

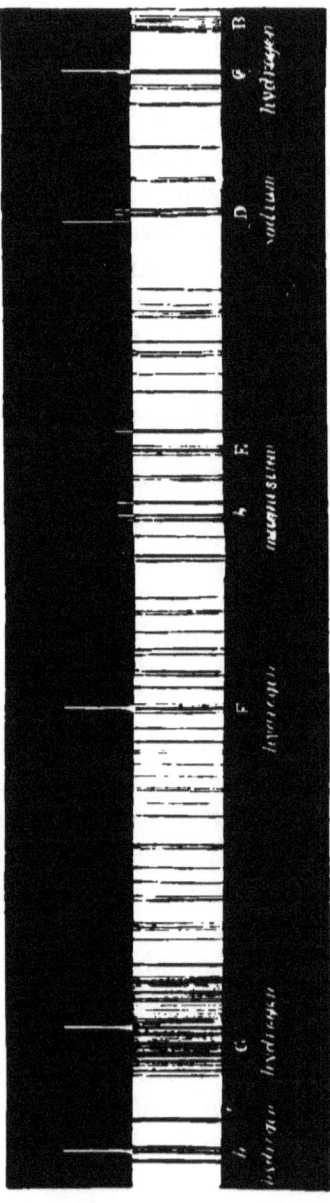

characteristic,—two almost equally strong dark lines across the spectrum, so close together as to be quite incapable of being separated from each other without the use of a telescope, or of a great number of prisms instead of one.

Now, Fraunhofer observed that in the flame of an ordinary tallow candle, when he tested it just as he had tested sunlight, there appeared a pair of bright lines, brighter than the rest of the otherwise continuous spectrum,—that there were no other lines in the spectrum but those two bright ones, and that they occupied, so far as his instrumental measurements enabled him to discover, precisely the same place in the spectrum as the two dark lines D of the solar spectrum. That is to say, the candlelight possessed in excess precisely one of those definite components in which sunlight had been found to be either wholly, or at least to a great extent, deficient.

No further action seems to have been taken with regard to this very remarkable coincidence, until Professor Miller of Cambridge, in 1849 or 1850, made a more exact experiment, with the view of comparing these yellow lines in the flame of a spirit-lamp with the dark lines of the solar spectrum, so as to test whether they are exactly coincident with one another or not. The result of his measurements was that the closeness of coincidence was so great that it was impossible, with his finest instruments, to find any divergence between them. The two bright lines exactly corresponded with the two dark ones as to refrangibility, and therefore also wave-length. It had been conclusively shown by Swan, that the two bright lines in the light of a candle are due to common salt, which pervades the air everywhere, and of which the very minutest trace is capable of producing this yellow light. It was then that Stokes at once took the additional step required, and explained that the glowing vapour, which is capable, when it is the source of light, of giving these definite bright lines, is itself, when used as an absorbing medium, capable of absorbing these and these only; and therefore that Miller's test of the exact coincidence of the bright and dark lines was a complete proof that there exists in the sun's atmosphere this vapour of sodium.

That occurred about 1850, and ever since that time the fact that sodium exists in an incandescent state in the sun's atmosphere, has been taught (as an experimentally ascertained truth) by Sir William Thomson and others. This was the birth of Spectrum Analysis, as applied to celestial objects.

It is curious to find that the deservedly celebrated Foucault had in 1849 made the same experiment in an even more convincing form than that which Miller had adopted. He found that the light, of what is called the electric arc, has in its spectrum these two bright lines; but that when he looked at sunlight through the electric arc and allowed the sunlight to come in so strongly as to overpower the electric light, then the electric light actually cut out the D lines from the solar spectrum more powerfully than if it had not been present. Although it was there giving out these lines strongly, it was not competent to fill up the wants in the solar spectrum, but actually made the deficiency more glaring than before. Then, to test whether it was really the case that this electric arc was absorbing these particular kinds of light, Foucault very ingeniously took advantage of the fact that the carbon points between which the electric arc is usually formed become incandescent and, reaching a higher temperature, are very much more brilliant than the arc itself. By means of a small mirror he reflected the white light from these carbon points through the electric arc, and found that whenever it passed through the arc, instead of getting brighter at those places, it had those very lines cut out of it; but whenever it passed beside the electric arc it had no deficiencies. Curiously enough, he seems to have derived no definitive conclusion from this.

Then, again, exactly the same statement was made in Sweden by Ångström in 1853. He says, as the result of experiment, that an incandescent gas gives out luminous rays of the same refrangibility as those which it absorbs.

Each one of these three thus completely made and recorded the discovery of the physical basis of spectrum analysis before 1854; but, of the three, Stokes alone made the application which really constitutes celestial chemistry. Fox Talbot had, long before, distinctly pointed out the use of the prismatic method for distinguishing terrestrial substances in a flame.

As I have already told you, Sir W. Thomson has, certainly ever since 1852 (probably a year or two sooner), regularly given in his public lectures in Glasgow University the statement that there is sodium vapour in the sun's atmosphere; and that, to find other constituents of solar and stellar atmospheres, all that is wanted is a comparison of the dark lines in their spectra with the bright lines in the incandescent vapours of various terrestrial substances.[1] But it was not till 1859 or 1860 that this was known generally, or was applied to any purpose further than to the mere recording of the existence of sodium in the vapours around the sun. I should like to read a quotation from some remarks I made a year or two ago to the Royal Society of Edinburgh upon this curious subject :[2]—

It is difficult now-a-days, when so many philosophers are engaged almost simultaneously at the same problem, to decide which of their successive steps in advance is that to which should really be attached the title of *discovery* (in its highest sense) as distinguished from mere *improvement* or *generalisation.* You have only to look at the recent voluminous discussions as to the discoverer of the Conservation of Energy, to see that critics may substantially agree as to facts and dates, while differing in the most extraor-

---

[1] President's *Address*, *Brit. Ass.* 1871. See Stokes, *Nature*, January 6, 1876. Thomson writes to me with reference to this (January 23, 1876) :—
' I never imagined that Stokes thought I was generalising too fast, or that *I* was generalising at all. I felt that I had learned the whole thing from him on a foundation of absolute certainty. . . . All I said in my Edinburgh Address on this matter is, I believe, irrefragable.'

[2] *Proceedings R.S.E.*, May 15, 1871.

dinary manner as to their deductions from them.[1] Some of these writers, no doubt, put themselves out of court at once by habitually attributing the gaseous laws of Boyle and Charles to Mariotte and Gay-Lussac. Men who persist in error on a point so absolutely clear as this, show themselves unfit to judge in any case of even a little more difficulty. Others, who strongly support the so-called claims of Mayer in the matter of Conservation of Energy, and who should (to be consistent) therefore far more strongly advocate the real claims of Talbot, Stokes, Ångström, Stewart, etc., to the discovery of spectrum analysis, are found to uphold Kirchhoff as alone entitled to any merit in the matter.

The question of priority just alluded to illustrates in a very curious way a singular and lamentable, though in one sense honourable, characteristic of many of the highest class of British scientific men ; *i.e.* their proneness to consider that what appears evident to them *cannot but* be known to others. I do not think that this can be called modesty ; it is rather a species of diffidence due to their consciousness that in general their accurate knowledge of the published developments of science is confined mainly to those branches to which they have specially devoted themselves. Their foreign competitors, on the other hand (especially the Germans), are often profoundly aware of all that has been done, or, at least, have some one at hand who is, and can thus, when a new idea occurs to them, at once recognise, or have determined for them, its novelty, and so instantly put it in type and secure it. Neither Stokes nor Thomson, in 1850, seems to have had the least idea that he had hit on anything new . . . —the matter appeared so simple and obvious to them—and, but for the fact that Thomson has given it in his public lectures ever since (at first giving it as something well known), they might have thus forfeited all claim to mention in connection with the discovery.

I went on to show how this lamentable state of things could easily be rendered impossible for the future, by the regular publication, at very short intervals, of a digest of all new advances in science.[2]

---

[1] Some frantic partisans of Papin, etc., deny almost all credit to Watt in the matter of the steam-engine! No further examples need be cited.

[2] [*Note to Third Edition.* To a great extent this desideratum is now

I come now to the question of what was done on this subject in 1858 and 1859. The first of these dates belongs to Balfour Stewart, and the second to Kirchhoff. Balfour Stewart treated the subject almost entirely from the point of view of what is called the Theory of Exchanges, and he demonstrated a very remarkable generalisation or extension of the law long before laid down by Prevost. Kirchhoff treated the subject from an, at first sight, somewhat higher theoretical point of view, and used reasoning considerably more complex and a good deal more mathematical; but in reality the fundamental point upon which the reasoning is based was precisely the same in both their investigations. What they established by their different processes was this,—the absolute equality of the radiating and absorbing powers of a substance for every definite ray. It was not merely what had been known to Leslie and others, that a body which is a good absorber of heat is also a good radiator of heat, with many other indefinite statements of that sort; but it was the precise limitation to each and every particular wave-length, and not only that, but something higher than that,—not merely to rays of a particular shade of colour, but also to rays polarised in particular planes. Stewart and Kirchhoff thus came to the conclusion that if you consider any one definite colour of light, and have it polarised in one definite direction, then a body which has a power of absorbing that, measured in any way whatever, will have an exactly equal power of radiating it, if measured according to the same units. So if we adopt the same units for radiating and for

supplied by the *Beiblätter zu den Annalen der Physik*; for which the whole scientific world is indebted to the disinterested labour of the Wiedemanns.]

absorbing power, in all bodies the measure of the absorbing power for any particular ray (strictly defined as I have just stated) is the measure of the radiating power for that same ray.

Stewart shows first, by very simple reasoning, that the absorption of a body at a given temperature must be equal to its radiation, for every given description of heat; and then he shows experimentally that a plate of rock-salt, which is an exceedingly bad absorber of heat, is also an exceedingly bad radiator of heat. Then he shows that a body is in general more opaque to radiations from another portion of the same body than it is to radiations from other bodies at the same temperature; in other words, if you measure how much of the heat radiated by a piece of hot glass is absorbed by rock-salt, and if you measure also how much of the radiation from an equally hot piece of rock-salt, instead of the glass, is absorbed by rock-salt, you find that rock-salt absorbs of the heat which is radiated by rock-salt a very much larger percentage than it absorbs of the heat which is radiated from glass at the same temperature; and this he showed to be true, with the change of a word or two, for mica, glass, and other substances. Then he showed also—and this is a very important addition—that a thick plate of rock-salt radiates more than a thin plate, being at the same temperature; and therefore it follows of course that the radiation from a hot body is radiation not merely from its surface, but also from layers under the surface, and in some substances it may be radiation from layers at a very great distance under the surface; so that radiation, like absorption, is not a mere surface phenomenon, but depends (when the substance is at all transparent) upon the depth or thickness of the absorbing or radiating stratum.

In order experimentally to show some of these results, though only in a qualitative not a quantitative manner, Stewart first tried a substance such as pottery ware, where you have a surface in some places white and in others black. If you look at such a piece of pottery ware by daylight, the reason why some markings on its surface are darker than others is simply that they absorb more of the incident light. These are portions of the body which absorb more than other portions, and therefore we should expect, if this law be true, and if it be capable of extension from heat rays to luminous rays, that on making the piece of pottery ware itself in turn the source of light,—making it hot enough to give off light,—then, as those portions which, when it was cold, appeared darkest, did so because they absorbed most, they should, when it is itself a source of light, appear brightest, because they ought to radiate most. That is an experiment which any of you can try very easily for himself with a piece of pottery which has a well-marked pattern on it. You will see, as soon as you have heated it to whiteness in the fire, taken it out and looked at it in the dark, a white pattern on a dark ground, instead of a dark pattern on a white ground. And it is very striking if, while thus looking at it, you suddenly flash daylight on it, when you see at once the inversion.

I can show to a few at a time, but not in a marked way at a distance, the same phenomenon, by taking a piece of platinum foil and writing letters on it with ink. When it is once heated there is a deposit, on the surface of the otherwise polished platinum foil, of oxide of iron which tarnishes the surface and makes it absorb considerably more light than a polished reflecting surface will do. We should expect, then, when this is heated (as I now heat it in a powerful but very slightly luminous flame), and becomes in turn the source of light, to see bright letters on a dark ground. The difference of brightness is not so marked in this case as in the last, but still those who are nearest to me will see the phenomenon distinctly enough.

But you will see another phenomenon still more startling on looking at the back of the heated foil instead of the front of it. You saw faint traces of bright letters on the dark ground when I turned the inked side to you, but when I turn the other side you see dark letters on a bright ground. Now, the reason why on the one side we have bright letters on a dark ground, while the other side of the same piece of metal shows dark letters on a white ground, is still more confirmatory of the result of Balfour Stewart's, which I have just stated, since these letters appear dark while at present cold, because they are absorbing more than the rest of the polished surface. They appear brighter than the polished surface when heated, because they radiate more; but just because they radiate more they must become colder, must be kept permanently colder than the rest of the foil, and therefore the parts at the back of the foil, behind those which are radiating most, remain permanently colder. This is made evident when we look at the side which is without any difference of surface, as we then see, by the relative amounts of brightness, a marked distinction between the parts which are hotter and those which are colder. This is a still more complete proof of Stewart's proposition.

Stewart extended his reasoning still further when he explained the behaviour of coloured glass when heated. If you look at a bright fire through a red glass (for instance); so long as the red glass is cold,—that is to say, is absorbing light but not radiating any,—it absorbs the green and lets the red through. That is why we call it a red glass, because it absorbs green and almost every ray but the red. When you put it into the fire and it has acquired exactly the same temperature as the coals, it shows no colour, and you cannot distinguish it from the coals. In fact, it is transmitting red light but is radiating green and other rays :—namely, those which it is capable of absorbing, and it is just making up by radiation for the amount which it is absorbing; and therefore the light which is coming through it from the coals behind is just as much strengthened both in quantity and quality by its radiation, as it is weakened by its absorption, and thus on the whole it comes to our eyes uncoloured. If you put in behind it, leaving it in the fire at a bright heat, a less hot coal, you will see it appears green, because then the glass is hotter than the background, and takes from the background less green than it gives out, and therefore the light which actually reaches the eye, partly through it and partly from it, is more green than that from the coal behind. Thus the coloured glass loses its colour when exactly at the temperature of the objects behind it, and takes the complementary colour when it is hotter than the objects behind it.

Kirchhoff gave a good many experimental illustrations of the relation between emission and absorption, of which I can allude to only two or three. The first was the very simple one of taking a bead of a transparent salt and heating it in a blow-pipe when supported in a loop of platinum wire. When the platinum wire and the bead were at the same high temperature, the platinum wire glowed bright, as we all know an incandescent wire does, but the bead of melted salt remained scarcely glowing at all. That is to say, this body, which is exceedingly transparent, and therefore a bad absorber, is also a bad radiator. On the other hand, the platinum wire is perfectly opaque; it is a good absorber, and therefore a good radiator.

But Kirchhoff carried his experiments after this at once to sunlight. Knowing that there is one particular definite red ray which is given out by the metal lithium when in a state of incandescent vapour, he noticed that there was no corresponding dark line in the solar spectrum. So he attempted with full success to make a new dark line in the solar spectrum by letting sunlight pass through a slit, and placing near the slit an otherwise slightly luminous flame (that of a Bunsen lamp) in which there was a large supply of lithium vapour, which caused it to give out light of one homogeneous red. When the sunlight came through it, the lithium vapour cut out that very red, and a new line was formed in the solar spectrum. As the sunlight was gradually weakened, the line gradually disappeared, though there was still a very visible supply of sunlight; and after a further weakening, the lithium line came brightly out on the darker background. Thus by regulating properly the intensity of the sunlight you may have at pleasure a dark line or a bright line, or no line at all, at this particular place in the spectrum. That was conclusive as to the possibility of producing these dark lines in new positions by the help of some incandescent gas.

But Kirchhoff also showed that if you take the direct radiation from a terrestrial body instead of—as in the last experiment I described—gradually checking or weakening the sunlight, if you had taken the light directly from a terrestrial source, then you could get the dark line only when the absorbing flame is colder than the source of light. In order to produce new dark lines in the sun's spectrum, we must use absorbing bodies which are colder than the sun. That of course presents no difficulty, because we cannot produce any terrestrial temperature which at all equals that of the sun ; but it comes to be a point of very great importance when we wish to produce these absorption bands in an otherwise continuous spectrum of artificial light, as, for instance, the light from an incandescent lime-ball. The temperature of the incandescent lime-ball is very low, compared with even the electric arc, and extremely low compared with the sun, but the light from it gives a perfectly continuous spectrum. If we try to produce in that spectrum the dark lines of lithium or sodium by the process just described, we find the process fail. Bright lines appear in place of dark ones. A Bunsen-lamp flame is not cold enough. In other words, if sodium be in the Bunsen flame, though it absorbs no doubt that particular orange light which comes from the lime-ball, yet in place of it it gives out a great deal more of the same kind, and therefore you have bright lines instead of dark ones. But if, instead of a Bunsen-lamp, you take an ordinary spirit-lamp, and put sodium vapour into its flame, you find you get your dark line in the spectrum of the lime-ball. Kirchhoff thus experimentally showed that for the production of an absorption-line (at least when the source and absorber are *both* behind the slit) it is necessary that the incandescent source should be at a higher temperature than the absorbing vapour.

We shall see that this is not only in itself a very important result, but that it is of the utmost importance when we come to interpret the spectrum we obtain from various portions of the sun's disc, and from various stars. It shows whether the radiating or absorbing matter in any of these cases is the hotter or the colder.

Finally, I may mention a discovery which was made almost simultaneously by Kirchhoff and by Stewart; the beautiful application, to absorbing bodies, of the polarisation of light. There are various transparent substances, which, although colourless, or but slightly coloured, nevertheless absorb all vibrations of light which take place in a particular direction. The simplest is a plate of tourmaline, cut parallel to the axis of the crystal. It is not yet absolutely certain whether the rays which it absorbs are those which vibrate parallel to the axis of the crystal or those whose vibrations are perpendicular to it, but that does not matter to our present purpose. Light which has passed through such a slice of crystal is vibrating in one definite direction only, and therefore is said to be polarised. As we have experimental proof that common light is subject to no such restriction, the portion of the incident light which has been absorbed must have been that whose vibrations were in the direction perpendicular to that of those which passed through, and therefore what was absorbed was also polarised light. Here, then, is a body which absorbs polarised light :—make it in its turn (by heating) a source of light, and it should then, if the proposition we are dealing with be universally true, radiate polarised light. Now the experiment has been made, and made with complete success. The light radiated by the red-hot crystal, if you view it against a dark background, is polarised; but if the background be itself as hot as the crystal (and be of non-polarising material), no polarisation is observed. The crystal transmits one part of the light unaltered, and though it stops the other half, it makes up for it by the light which it radiates.

Before I go further with the reasoning on this part of the subject, I must make a slight digression as to our knowledge, or rather our reasons for conviction, of the identity of radiant heat and light. I must, in fact, show how we satisfy ourselves that there is no more difference between radiant heat and light, or even the so-called actinic sun rays, etc., than between waves of sound or waves of water of different lengths. The precise nature of the vibration which constitutes a wave of light, does not matter to this question at all.

We all know that sound-waves differ practically from water-waves. In the case of sound-waves the particles of air are vibrating back and forward in the direction in which the sound travels; but in the case of waves on water, the particles are moving partly up and down and partly back and forwards, so that near the surface of the water each particle is describing almost a circle.

In the case of luminous waves, all we know is, that whatever be the precise nature of the vibration, its direction is transverse to the direction in which the light is moving. Now, the proof that radiant heat and light are the same, or only variations of the same thing, is to be arrived at by comparing their properties in a great number of different ways. I cannot enter very deeply into this, but I can at all events mention several of these properties, and show how conclusive the evidence is for the identity in question.

First of all, it is to be considered they both move in straight lines. The radiant heat from the sun goes along with the light from the sun, and when you shut one off,—put an opaque screen so as to intercept the one,—the other is intercepted at the same time. In the case of a solar eclipse, you have part of the sun's heat as long as you can see the smallest portion of the sun's disc. The instant the last portion of the disc is obscured, the heat disappears with the light. That shows that the heat and light take not only the same course, but also the same time to come to us. If the one lagged ever so little behind the other,—if the heat disappeared sooner than the light, or the light sooner than the heat,—it would show that though they both moved in straight lines, the one moved faster than the other; but the result of observation is that we find, so far as our most delicate measurements show, that heat and light pass from the moon to the earth, *i.e.* over a space of a quarter of a million miles, in sensibly the same time. Therefore we have the proposition that radiant heat moves at the rate of about 186,000 miles per second, because that is the velocity of light. Thus, even our very first analogy between them seems to be almost convincing as to their identity.

Then we have, as you all know, when you use either a burning mirror or a burning lens for the purpose of condensing the sun's heat into a focus, to adjust this by taking advantage of the sun's light. You form an image by means of the light, and then you find the heat rays concentrated at the same point. That is to say, the laws of reflection and refraction are precisely the same for light and radiant heat.

Then again, I dare say you all know—for it was shown early in this century as the true explanation of phenomena known to Newton, and before him—that two sets of rays of light can be made so to interfere with one another as to produce darkness. This experiment is conclusive[1] as to the non-materiality of light, and shows that light must be something of the nature of a vibration of some kind, so that two opposite motions meeting in one place, or rather simultaneously affecting one part of a medium, may produce simple rest or non-existence of motion. If we test this with sunlight, as was done by Fizeau and Foucault, we find that precisely at the places where the sunlight has disappeared, at these same places the sun's heat has disappeared. Radiant heat, therefore, has, as light has, the property of interference; that is to say, two portions of either are, in certain circumstances, capable of mutually destroying one another.

Another very striking analogy between them is furnished by absorption. Let us take the case of light first. If I take a number of pieces of the same blue glass, light which has passed through one of these is capable of passing in greater part or percentage through the next, and what has been sifted through two of them will in still greater percentage pass through the third, and so on. Precisely the same thing holds with reference to radiant heat. What we call colourless glass happens to be extremely opaque to radiant heat, especially to the lower forms; but if you force, by using a powerful source, a considerable beam of radiant heat through a single plate of window glass, you will find that though

[1] Compare foot-note to p. 66. The applicability of the word 'conclusive' is matter of opinion.

the window glass is exceedingly opaque to such heat in general, what you can force through it will pass in very much increased percentage through the second plate, and in still greater percentage through the third, and so on. And, just as Melloni used the word *Thermochrose*, we may say that a pane of window glass, which is colourless or almost so as regards light, would be regarded as coloured by larger beings than ourselves: beings with such very coarse-grained optical apparatus as to have the sense of light produced in them by such waves only as to our senses produce radiant heat. Such creatures would speak of our most transparent glass as being exceedingly opaque, while they would speak of rock-salt as being transparent, for it is found to transmit heat almost as freely as it does light.

There are various other analogies, such as, for instance, that the intensity of light from any source varies inversely as the square of the distance. The same thing is true of radiant heat. That, however, is necessarily true of all emanations in straight lines from a centre when there is no absorption, so that it does not strengthen the argument. It is, in fact, merely a consequence of the geometrical truth that the surface of a sphere is as the square of its radius.

Then again—and this is perhaps the grand proof—we have the discovery by Principal Forbes of the polarisation of heat. You can polarise radiant heat as you can light, and this is the most conclusive argument; one which, taken with what I have just told you, leaves no possibility of escape from the conclusion that the difference between radiant heat and light is simply the difference between a low note and a high one. Therefore, in reasoning upon radiation, it is quite indifferent whether we speak of radiant heat or radiant light, or even higher waves which are invisible to the eye except through fluorescence. So we need speak of nothing but radiation, under which we suppose them all included.

Now comes this question—By what marks can you distinguish one particular radiation from every other? Well, you can do that just as you can perfectly define any particular sound. You can define a sound if you are told three things about it,—its intensity, its pitch, and its quality. Well, the quality has of late been shown by the beautiful analytic and synthetic methods of Helmholtz to depend upon the admixture of other sounds (harmonics) with the primitive sound; so suppose you take the simplest quality of all,—that which has no admixture,—then a musical sound or note is completely defined if we know its intensity, if we know its pitch, and if we know its quality to be the simplest. Conversely, if you have any disturbance of the air which has that intensity, and that pitch, and the simplest quality, it will be that same sound. That, then, is our mark by which we detect a particular kind of motion of the air.

Now we have precisely the same sort of marks by which we can distinguish a particular kind of radiation. Take the simplest form,—that is, the simplest quality,—there are other three things to be attended to. The first is the intensity; the second is the wave-length or colour, what corresponds to the pitch; and the third, how it is polarised, or whether it is polarised or not. If these be attended to, and if you were to specify them for any one radiation, and if any other radiation whatever satisfied the same conditions, then it must necessarily be the same radiation. That is the stamp of equivalence between the two.

And now we come to the question how we prove that the radiation from a body must be equal to the absorption by the same body under similar circumstances. We have seen how we can test the equality or identity of two radiations, and now it remains, having that preliminary settled, to apply reasoning to see why the absorbing and radiating powers are necessarily equal.

The best way we can do it is by applying reasoning very similar to Carnot's, but in this case it happens to be capable of being applied even more simply. Suppose we have a space, the walls of which are either perfect reflectors or are always kept at a definite temperature. It is an experimental fact that bodies (whatever they be) which have been long enough kept in either kind of enclosure will at last acquire precisely the temperature of the enclosure. That is a fact which has been ascertained over and over again. We say, in fact, that bodies are at the same temperature when neither parts with heat to the other, when there is on the whole no transference of heat from the one to the other when they are placed in contact. Suppose one of these bodies more capable of absorbing than it is capable of radiating. That body would be constantly taking in more heat than it was giving out, and therefore though the other bodies would of course absorb the heat which was given out by it, they would necessarily cool, because they would get back from it less than they gave to it. It would be getting hotter at the expense of all the other bodies inside the enclosure.

So far then the reasoning appears, at least at first sight, to be nearly complete; but it is not, because we have been taking the radiations as a whole. Suppose we have then inside this enclosure, between two of the bodies, some body which we may call a screen, and which shall allow to pass a perfectly definite kind of radiation and that alone, completely reflecting everything else. Then whatever radiation passes from the first towards the second body must pass through the screen, and will be, therefore, of this definite kind only. It will be partly absorbed and partly rejected by the second; but if there is to be a constant equality of temperature maintained—and that is our fundamental proposition—the fraction of the amount of this definite kind of radiation given out by the first and absorbed by the second must be exactly equal to the fraction of that given out by the second, and absorbed by the first, else one of the bodies would rise in temperature at the expense of the other, and that, you know, is impossible. Such a screen, in fact, while (with regard to the particular radiation in question) it forms one of three bodies in the enclosure, virtually (for all other radiations) makes it into two separate enclosures.

You will notice that our reasoning is in reality based upon Carnot's:—the same which led to the second law of thermo-dynamics, because it is founded on this principle, that without expenditure of work we cannot cool down a body below the temperature of all the surrounding bodies. If a body is at the same temperature as the surrounding bodies, we cannot use its heat to do work. In fact, we require to spend work to make this body colder. Now, if it could make itself colder by radiating away more than it absorbs, then we should have an enclosure containing two kinds of bodies, one of which heated itself at the expense of the other, so that work could be got from bodies all originally at the same temperature, and kept in an enclosure which is throughout constantly at that temperature. [But, just as we have seen that Carnot's principle is only true in the statistical sense, and would not hold if we could deal with individual particles of matter, so this assertion of the equality of radiating and absorbing powers is true in a similar statistical sense only.]

A word or two about the differences between different bodies. There are some bodies which absorb every radiation which falls upon them. These are such bodies as lamp-black, and we may call them black bodies in general. Now, as a black body is capable, by definition, of absorbing every kind of radiation which falls upon it, so it must, by applying to it the proof we have given, be a body which when heated must give off every kind of radiation. There can be nothing wanting,—no dark lines in its spectrum when incandescent,—because as it is capable of absorbing everything, it is capable of radiating and will radiate everything.

The next class of bodies we may call transparent bodies. A transparent body is a body which absorbs nothing at all. If we had a perfectly transparent body it would absorb nothing, and therefore would radiate nothing if you made it hot, so that the body could not be seen either by itself stopping a certain amount of the light which falls upon it, nor could it be seen by making itself a source of light, because, in consequence of its not being able to absorb, it would not be able to radiate. It might, of course, be seen in virtue of its displacing, or distorting, the images of other bodies as seen through it.

Then we have, finally, a class of bodies which may be

distinguished from the other two as those into which heat cannot be absorbed at all, because it never penetrates the surface,—those which are perfectly reflecting bodies.

We have, then, black bodies, transparent bodies, and reflecting bodies. We have none of them in perfection, but we may take lamp-black as an example of the first; rock-salt for the second; and a polished metal, such as silver, for the third. None of these is perfect; but they are all approximations, and are as good approximations as we find in Nature to the mathematical ideas of rigid bodies and perfect fluids, and sufficiently near approximations to enable us to deduce from our reasoning on them valuable explanations of physical phenomena.

Let us suppose for a moment that we have two of these bodies inside a perfectly reflecting enclosure. Suppose one to be a black body and the other a transparent body,—only let it be imperfectly transparent. Then the black body takes up absolutely any radiation which may come to it, and sends out absolutely all kinds of radiation; but the transparent body is incapable of absorbing any but one kind of radiation, let us say. It will go on absorbing that one kind of radiation from the black body, but the black body gets back by reflection and transmission, all except that one kind of radiation, and therefore that must be the one kind of radiation which the transparent body can give out: and it must give it out,—being at the same temperature as the other body in the enclosure,—it must give it out precisely at the same rate at which it absorbs it; and thus we have from a somewhat varied point of view another demonstration of the same principle. I shall endeavour in my next lecture to illustrate these theoretical conclusions by experiments.

LECTURE IX.

SPECTRUM ANALYSIS.

Spectrum of incandescent black body; of incandescent gas or vapour. Absorption by vapour of parts of spectrum of incandescent black body. Application to sunlight, and starlight. Solar spots and protuberances. Period of life of various stars. Fluorescence.

THE point at which I had arrived in my last lecture was the practical results of the independent discoveries—because we can call them no less—of Foucault, Stokes, Ångström, Balfour Stewart, Kirchhoff, and others, with regard to the equality of the radiating and absorbing powers of any one body for any definite ray of heat or light. I explained very fully in that lecture how we can test, by separating them from one another, all the different forms of radiation that proceed from any particular incandescent body, and so discover whether any are wanting. It only now remains that I try these experiments with the help of a galvanic battery, of which I have a pretty powerful specimen down-stairs, connected by wires with the electric lamp here.

In an ordinary galvanic battery, if you have only a moderately great number of cells, say 100, no electricity will pass between the terminals until you bring them into contact; but if after bringing them into contact you then separate them, a spark will follow, and heat the air between them so much that, if the battery be powerful enough, we may have a steady current of electricity passing between the two, and keeping the intervening air in a state of incandescence.

Now everything in the path of this portion of the current is so intensely hot that any ordinary metals such as copper, would be melted at once, or at least in a very short time, if placed in it; and therefore the substance we employ for the poles of the battery is gas coke, the hard deposit of carbon found in gas retorts. Bars of this are cut and connected with the poles of the battery. When the voltaic arc is passing in hot air between the two poles, the ends of these bars become vividly incandescent, and you have therefore two very hot black bodies, and between them a hot semi-transparent body. Now, you will remember from my last lecture that a black body is black because it absorbs all kinds of light which fall upon it. A transparent body is transparent because it absorbs little of the light which falls upon it.

Therefore, if our proposition be true,—and we know it must be, in the same sense at least as is the second law of the dynamical theory of heat,—it will follow that if you make the black body incandescent it will give out all kinds of radiations, just as it is capable of absorbing all kinds; and the gaseous or semi-transparent body, which is capable of absorbing only few kinds of light or radiations, will be capable, when self-luminous, of giving out only the same few. The contrast between the two will be well seen by adopting the optical method I described in my last lecture. We separate from each other the various kinds of radiation given by the two bodies simultaneously.

The radiation which you see just now [throwing the spectrum of light from one of the carbon points on the screen] is mainly, almost wholly, from one of the hot carbon points, and you see that in its spectrum there is no discontinuity. You have every colour, or rather wave-length, of visible light, from the lowest red which the eye can see, up to the highest violet which the eye can see. There are irregularities of brightness in the spectrum, but these are due to the fact that we have the glowing gas as well as the luminous black body. But if I pass to the spectrum of the glowing gas by altering (as I now do) the position of the carbon points inside the electric lamp, you now see in the dark interval between the two continuous spectra on the screen, each belonging to one of the carbon points, bright lines showing definite kinds of radiation and those only, while the black bodies give you every possible variety of tint, as it were, from the lowest up to the highest visible. This glowing gas, which is the arc between the two poles, gives you only certain definite kinds of light. At present it is very difficult to tell, without careful measurement, to what particular vapour each of these rays belongs, because the composition of the glowing gas between the poles depends for the moment entirely, or almost entirely, upon the impurities in the carbon points, which have been vaporised by the intense heat; and, therefore, though I can at once see, partly from its position and colour, but much more definitely from its brightness, that the orange-coloured ray belongs to the metal sodium, I could not, without careful measurement, tell what other substances are incandescent in the spark there to produce the other bright lines. But if I were to introduce some substance into the arc (placing it upon the comparatively large upper surface of the lower pole), I should be able to see what lines it produces, whether by new lines appearing or by the strengthening of those already present. To illustrate this, I shall take a small piece of metallic sodium, and render it incandescent upon one of the carbon points ; then, as it is highly volatile, the space between the two carbon points will immediately be filled principally with vapour of sodium. You see at once that the orange line, to which I have already called your attention, is very greatly intensified—the others being but little affected, though, if anything, rather weaker than before. That depends upon the fact that the sodium vapour offers much less resistance to the voltaic arc than does air. The arc is both longer and at a lower temperature than it was before I introduced the sodium.

To show the testing nature of this mode of discriminating between different substances, I take a small portion of a metal which was discovered by the help of the spectroscope almost immediately after this method of observation was brought into practical use. You see that, in addition to the feeble bands which cross the comparatively dark space between the continuous spectra of the carbon poles, we have a new one of great intensity and of an exquisite green colour. This is characteristic of the vapour of the metal Thallium (closely allied to lead in many of its physical and chemical properties), a small portion of which I had placed upon the lower carbon.

The final experiment I have to show in this connection is the converse of this. We are now going to take as the source of light one of the carbon points, an incandescent black body which gives us all kinds of radiations, from the lowest to the highest, and we are going to make the sodium vapour, whose particular kind of radiation we have already studied, the absorbing body, and see what part of the continuous spectrum of the incandescent black body it cuts out or refuses to allow to pass. And in this experiment of course rests the definite proof, so far as one single case of experiment can give a definite proof, that the absorption and radiation are exactly equivalent to one another in every particular glowing gas.

I place near the slit of the electric lamp a powerful Bunsen burner, into which my assistant will introduce a pellet of metallic sodium in a little iron spoon. You see first the combustion of the naphtha in which the sodium was kept (to prevent its oxidation), then in a few moments you have an excessively bright monochromatic flame, whose light is due almost entirely to the incandescent sodium vapour. You see the weird expression of one another's countenances as this snap-dragon flame becomes more and more intense. I interpose a sheet of pasteboard to prevent its direct light from falling on the screen where the spectrum of the carbon point is for the moment seen continuous. I now move the Bunsen lamp slightly, so that the carbon point must shine through the sodium flame, and you see at once a dark, almost perfectly black, band cut out of the otherwise continuous spectrum, just as if a pencil or other opaque body had been interposed. You see it is exactly a prolongation of the orange band of sodium still furnished by the voltaic arc, and you see that it appears and disappears exactly as I put the lamp in front of the slit or withdraw it.

It would be easy to extend a series of experiments of this kind, but there would be a very great deal of sameness about them, because all we could do would be to show over and over again that certain bodies when incandescent give perfectly definite kinds of light; and that the same bodies when incandescent, but sufficiently colder than the carbon pole of the electric lamp, cut out from the otherwise continuous spectrum of the carbon pole precisely the kind of light they give out when they are themselves made the source of light. But in order to convert this rude experiment into a perfectly definite physical method of measurement and proof, it is necessary to take more refined means of comparison than the methods I have just used. First of all, it is important to make the slit an extremely narrow one. I was obliged to make the slit moderately wide that you might see the various coloured images of it; but in order that we may have a thoroughly trustworthy measurement, I must make the slit extremely narrow, and then we shall have perfectly sharp definite lines, only of the breadth of the slit itself, showing these colours. You see them now perhaps half-an-inch or one-third of an inch broad; but it is perfectly possible and necessary for exact physical measurement, to make them excessively narrow, and to measure with the most extreme care their relative positions with regard to one another. Another necessary detail is to place the prism, or prisms, exactly in the position of minimum deviation as it is called—in which case the rays make equal angles with the surfaces of the prism at which they enter and escape. Then and only then are we entitled to conclude that there is absolute coincidence between the dark absorption lines and the bright lines due to the same incandescent vapour, according as it is employed to absorb or to radiate. When this is to be carried out with the utmost perfection attainable in modern optics, we employ, not the method of projection upon a screen, which I have used just now, but a far more delicate method, invented by Fraunhofer, in which the rays are received by the object-glass of a telescope, so that in the air, at its focus, an image of the spectrum may be formed. This may be examined by means of an eye-piece, as powerful as we choose, so that we may separate the different kinds of light radiated by a glowing gas, by telescopic power as well as by increasing the number of prisms. By using telescopes more and more powerful, and greater numbers of prisms, as you can easily conceive, this method will enable us to measure with the utmost nicety, to any degree of approximation that may be desired, the relative distances between the various lines of the spectrum. So we have the means, if we apply these refined methods to light from celestial sources, and also to that from known terrestrial sources, of determining whether the different radiations and absorptions observed belong to precisely the same wave-lengths, that is, have precisely the same positions in the spectrum; and therefore we have as complete physical proof as it is possible to desire of the presence (somewhere or other in the path of the light which comes to us from a celestial body) of the incandescent vapour of a particular known terrestrial substance.

This, then, is the basis of spectrum analysis as applied to problems connected with the physical universe. I shall now say a few words about the results of the application of this method of investigation to the sun, stars, nebulæ, and comets. The literature of this subject has become very extensive considering how new it is. Seeing that the subject is barely fourteen years old in its definite applications, it is astonishing to find that it already fills many volumes of special treatises and a host of scattered papers, and still more to find that a great deal of what is there contained is thoroughly popular and yet thoroughly trustworthy.

On a point of this kind, therefore, most of you by a little reading can acquire at least as much information as I have to give you. Therefore, while I must take some notice of it, I need not at all dilate upon it, though it is a very important and interesting part of our subject.

First of all, let us consider what we do see when we treat sunlight as it comes to us from the sun, as a whole, that is without specifying any particular portion of the surface. Take a beam of sunlight and subject it to the same scrutiny which we have employed to-day upon the light of the electric arc and the carbon points. It is impossible in any coloured diagram to represent accurately the solar spectrum, and therefore no graphical delineation will at all supply the place of an actual examination of the phenomena. Nor will a verbal description; so I shall be very brief. We find at once that the solar spectrum is crossed over by an enormous number of black lines, perpendicular to its length; precisely as the black line lay which you saw a little ago across the otherwise complete spectrum from which it had cut out a portion. We arrive therefore at the conclusion that the sunlight must have come originally from some black body, or opaque body, which is intensely self-luminous, and which may be either in a solid or in a liquid state,—possibly even in the state of extremely compressed gas. However this may be, the source of light in the sun, whatever it is, must, in so far as we can see, give off all kinds of radiations, so it is practically a black body. These black lines, or gaps, in what would otherwise be a continuous spectrum, must therefore be due to absorption by vapours (self-luminous or not) which are somewhere in the path by which these sun rays arrive at our earth.

Now, the source of a part of these lines has been known for a very long time,—since Sir David Brewster's early days, in fact, for he discovered that they were due to absorption by the earth's atmosphere. We know the earth's atmosphere does absorb a great deal of sunlight. The rising sun, when we see it obscured by vapours, is by no means comparable with the sun in the zenith; but that is mainly a kind of absorption which would be given by neutral tinted coloured glass, which would tone down the various rays in only slightly different proportions, very much, in fact, as they are toned down by reflection, as when you see an image of the sun in a pool. The reflected sun is very much less bright than the direct, and after two or three reflections from a glass surface may be looked at without injury to the eye. But here the effect is a mere general weakening of the light, there is no special or selective absorption. The atmosphere might merely have weakened the various kinds of sunlight in some such nearly constant proportion, but Sir David Brewster found it did more than that. He found that when you compare the solar spectrum when the sun is high with that of the same sun when it is rising or setting, there are a great many more lines crossing it in the latter than in the former case; and he concluded that, as the only difference of circumstances between the two cases is that the same rays had to pass through a much longer extent of the earth's atmosphere (and especially through the dense part of it), at sunrise or sunset, than when the sun is high, therefore these new lines at least are due to absorption by the air, or by aqueous or other vapour in the air.

It is possible, by that very simple comparison of the spectrum of the sun at rising with the spectrum of the sun at mid-day, to classify the missing rays, and say there are some whose absence is obviously due to the earth's atmosphere; the remaining ones we cannot account for by anything terrestrial,—we must go either to the space between us and the sun, or to the sun's atmosphere for the explanation of their cause.

Now, it is obvious that if the absorption were due not to the sun's atmosphere or to the earth's atmosphere, but to some other medium between us and the sun, that medium would treat the light of all the other stars just as it treats the light of the sun; and therefore if these lines in the solar spectrum which are not accounted for by the earth's atmosphere can be accounted for by anything in space, all stars should have spectra containing the same dark lines as are found in that of the sun.

Now that has been found to be by no means the case. Many stars have spectra totally different from that of the sun, as well as from one another. Therefore the spectrum given by any particular sun or star is due mainly to its own atmosphere of incandescent vapour, and we can thus study the chemical composition of the atmosphere of that sun by simply finding what terrestrial substances, put into a Bunsen flame, or rendered incandescent by electricity, will produce bright lines in its spectrum corresponding to the dark lines we find in the spectrum of the star. Here is a small portion of the grand drawing made by Ångström, a mere fragment of his map of the solar spectrum,—not above one-thirtieth of the whole he has depicted. The numbers above indicate the wave-length, in fractions of a millimetre. Thus the three conspicuous green lines of magnesium, forming the group called *b* by Fraunhofer (see diagram, p. 192), are seen here to have wave-lengths of $0^{mm}\cdot0005167$, $0^{mm}\cdot0005172$, and $0^{mm}\cdot0005183$ respectively. This portion, as you see, contains a number

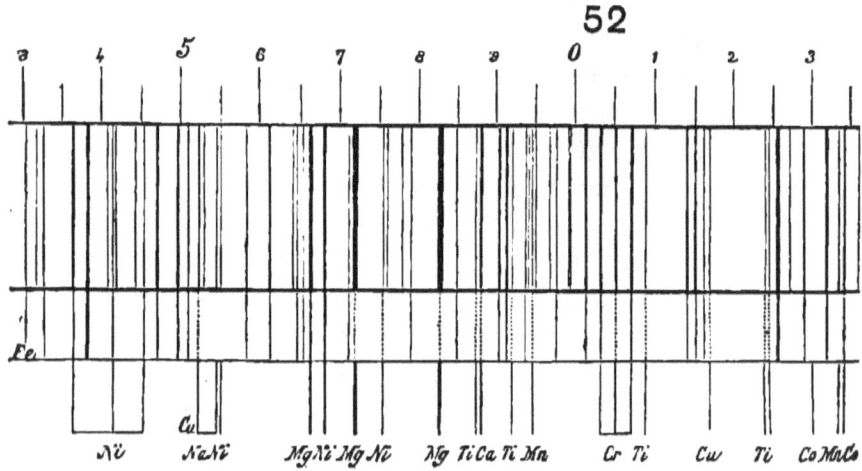

of dark lines. Well, when you pass sunlight through one-half of the slit of the spectroscope, and light from incandescent materials (in the electric arc or in an induction spark, or even a Bunsen lamp) through the remaining half, and examine them through the same train of prisms, you get two spectra as here represented, the one of sunlight and the other of the terrestrial substance, spread out side by side. Any two rays which, in passing through a very long series of prisms, undergo exactly the same treatment, must be of the same refrangibility; and therefore, by what I have just explained to you, due to the same definite substance.

In the band just below the portion denoting the solar spectrum, all the full lines represent lines which are actually observed in the spectrum of metallic iron. Looking up, you see the exact coincidence of each with a corresponding line in the solar spectrum. You see there are about thirty coincidences even in this small part of the solar spectrum; and so throughout the spectrum the number of coincidences between actual bright lines given by incandescent iron, and absorption lines in the solar spectrum, may amount to several hundreds. By recording both spectra photographically, it appears probable, from some recent experiments, that these hundreds of observed coincidences may in a short time become thousands. Now, as Kirchhoff has shown, even if there were not an absolutely ascertained coincidence in any one of these cases,—if it were only so near a coincidence that we could not be perfectly certain, by means of our instruments, that it was an exact coincidence,—still, looking at the question from the point of view of the theory of probabilities, the chances of iron's not existing as an absorbing medium in the sun's atmosphere, as estimated by a person who has seen even a moderate series of at least approximate coincidences, would be represented by one against a number which I cannot pretend to understand, but which contains some thirty-five places of figures. You can see then what extraordinary sort of probability there is that iron is there; and when I say that that probability was derived from only a comparatively few coincidences in the solar spectrum, how enormously greater would it not be were we to take account of all now known. And not only this. Lines which, in the iron spectrum, are strong,—are correspondingly strong in the solar spectrum—to every grade of nicety. So far, then, iron must be in large quantities in the sun's atmosphere. We find also that nickel must exist there. Every bright line shown by incandescent vapour of specimens of nickel in our laboratories (whether these specimens be terrestrial or cosmical, *i.e.* meteoric) corresponds to a dark line in the solar spectrum. Not only so, but the character of each bright line and the corresponding absorption line is the same. Very bright lines correspond with very dark ones, broad lines with broad, narrow with narrow, double with double. Some lines appear to be given by two different substances, as iron and nickel, for instance. This is probably, in the great majority of cases, due to slight impurities of the specimens tried. Various other substances are shown in this small portion of the spectrum,— magnesium, manganese, cobalt, chromium, sodium, titanium, and calcium. The number of titanium lines has been shown by Thalén to be very much in excess of even the enormous number of iron lines I have mentioned.

So far then this has been a question of the spectrum of light taken from the whole surface of the sun; but it becomes an exceedingly curious question, Are there local differences in the light from that surface, and if so, what are they? when we reflect that there are such things as sun-spots, and also when we think of those peculiar red flames, as they used to be called, which are seen round the dark body of the moon during a total eclipse. It becomes an exceedingly curious question what we shall get if we take sunlight from a limited portion of the sun's surface, as we can do by using a telescope lens of long focus. We form by means of it an image of the sun of an inch or so in diameter, and place the slit of our spectroscope successively on various parts of that image.

It was to be expected that some very important additional information would thus be obtained. Now, such information you can quite easily procure for yourselves by reading works like that of Lockyer, but I may just very briefly indicate its nature. In the first place, we

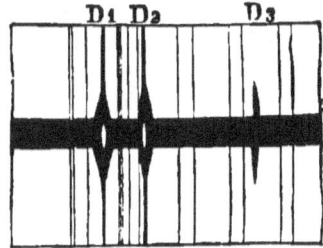

find that from sun-spots in general we have those absorption lines a little thicker and darker than from sunlight as a whole, so that it appears that there is associated with the sun-spot something which produces an excess of absorption. There is a more powerfully absorbing medium at the place where the sun-spot appears than at the places where faculæ or bright spots appear. In the particular spot, a portion of whose spectrum is here figured, the lines (D_1 and D_2) of sodium appear not only broadened over the spot, but *reversed:* —*i.e.* bright instead of dark :—just over the middle of the spot.

Then, when we come to examine the red flames or

prominences, we find that in general their spectra consist simply of bright lines. Such then is the spectrum of part at least of the gaseous matter which surrounds the sun, and it is the upper portion of the absorbing medium which cuts out these black lines from what would otherwise be a continuous spectrum, and you easily trace what lines it does cut out. For instance, here is a dark line (C) in the red [see diagram, p. 192, which shows, as through the same slit, the spectra of

the sun and of a prominence], which is due to hydrogen gas. Well, we find these red flames owe their redness to the particular colour of this line of hydrogen. So this bright red line is one of the main features of the prominences. Then we find a yellow line very nearly coincident, as you see, with the lines of sodium. Nobody as yet knows what is the chemical substance which produces this particular line. It corresponds to no absorption line usually found in the sun's spectrum (though you observe a trace of it in the spot spectrum which I last showed you), and therefore it must be due to a substance in a peculiar condition capable of radiating, but of having its absorption made up for,—some substance which possibly we may not yet know. Possibly it may not be a terrestrial substance at all. But it occurs here, very nearly giving a coincidence with sodium; but its light is not only more refrangible, but it wants the distinctive property which sodium has of giving a double line. Then we find several other lines, including two—or I may say three—more, due to hydrogen; so that the spectrum of these flames consists mainly of the spectrum of incandescent hydrogen gas. Here is another drawing of a small portion of the spectra of the sun and a prominence, which shows the exact coincidence of the bright and dark lines.

Suppose now we had a telescope to which the spectroscope could be adjusted: on looking at a red prominence without the spectroscope we should see one image, but it would be an image which consisted partly of the red, partly of the green, partly of the blue, partly of the violet rays of hydrogen; but if we combine telescope and spectroscope, the combination would enable us to separate from each other, along the line of dispersion, the various colours; and the edge of the sun would be treated in the same way. All its colours would be spread out from one another, but they would be spread out at a disadvantage compared with the colour of a monochromatic line. Because however far you separate one such line from another, you do not weaken either. They remain, except in so far as reflection from the surfaces of the prisms, and absorption within the prisms, weaken them, as strong as ever. But if you take a corresponding portion of sunlight, then, since it gives practically a continuous spectrum, you spread it uniformly over as long a space as you choose. So by the aid of this property, as the solar spectrum is practically continuous, except where there are interceptions of light, you can spread it out, and thus weaken it throughout as much as you please; whereas the other spectrum consists of perfectly definite bright lines, which you may spread as far apart from one another as you please, but which you cannot individually weaken. Hence, however strong be the glare of sunlight, sufficient dispersive power will enable us in fine weather to examine the spectrum of the red flames.

This is perfectly analogous to the observing stars by daylight, which, you are aware, is done in every fixed observatory by means of a good telescope. It is simply because the diffused light of the sky allows itself to be weakened farther and farther as we spread it over a larger and larger image, while the light of the star always comes from the same definite point; because no one has yet made a telescope showing a star's disc (except as a delusive appearance due to diffraction), so that, magnify it as you please, its light comes from the same definite point. So it remains of the same brightness, while the background may be made as dark as you please by spreading it out. In that way, by combining the spectroscope with the telescope, and widening, or altogether dispensing with the slit, it is possible to study the phenomena of these red flames, and, in fact, the whole behaviour of gaseous matters round the edge of the sun's disc, without waiting for a total eclipse. This is an extremely beautiful adaptation of means first made theoretically by Lockyer, and afterwards by Janssen, but brought into practice nearly simultaneously by the two astronomers.

Here is the result as applied to a particular portion of the sun's circumference. The body of the sun we

will suppose to be under that picture. These are simply eruptions of glowing gas from the sun's apparent surface. On the same scale there would be another image, a green image, situated almost at the end of the room; then a long way beyond, an indigo, and finally a violet

one. But we have by means of the prisms separated that particular image from the others, and thus we have here a monochromatic representation of what is above the surface of the sun, in so far at least as incandescent hydrogen gas is involved. When I point out that the change from the first figure to the second took place in the course of a few minutes, you will see what exceedingly rapid changes are going on in these self-luminous clouds; and when I further tell you that the height of this prominence, which is a stream of hydrogen rushing violently up from a rent in the surface of the sun, is something like 70,000 miles, you will see on what a stupendous scale, and with what tremendous velocities, these phenomena are constantly taking place.

So far then for the sun. When we compare the spectra of different stars with that of the sun, we come to some very curious conclusions. We find four classes of spectra, as a rule, among the different fixed stars which have seemed of importance enough to be separately examined. The first class of spectra are those of *white* or *blue* stars. You see an admirable example in Vega, and another in Sirius, or the dog-star. All these white stars have this characteristic, that they have an almost continuous spectrum with few and broad dark lines crossing it, and these few for the most part lines of hydrogen. These stars are in all probability at a considerably higher temperature than the sun; and their atmospheres are in even more violent agitation than is that of the sun. Then you come to the class of *yellow* stars, of which our sun is an example. In their spectra you have many more dark lines than in those of the white stars, but you have nothing of the nature of nebulous bands crossing the spectrum, such as you find in the third class; still less have you certain curious zones of shaded lines which you have in the fourth class of stars. This classification seems to point out the period of life, or phase of life, of each particular star or sun. When it is first formed, by the impact of enormous quantities of matter coming together by gravitation, you have the very nearly continuous spectrum of a glowing white hot liquid or solid body (or, it may be, dense gas), the sole, or nearly sole, absorbent being gaseous hydrogen in comparatively small quantity, and the spectrum having therefore few absorption lines. As it gradually cools, more and more of those gases surrounding its glowing surface become absorbent, and so you have a greater number and variety of lines. Then, as it still further cools, you have those nebulous bands which seem to indicate the presence of compound substances; which could not exist in the first two classes, because there the temperature is so high as to produce dissociation. Still further complexity of compounds will be found in the atmospheres of the fourth class. But sometimes, as in the case of temporary stars, a spectrum of the fourth class is suddenly crossed by the bright lines of hydrogen—showing either a last effort at the discharging of red flames, or a flicker due to some last chance impact of meteoric matter. So that we can study, as it were, not the succession of phases of life in any one particular star, but different simultaneous phases in many : we can study some stars, as it were starting into life, others getting older, others older and older; and we occasionally find a most remarkable circumstance happening with a star that has practically died out,—a star which is scarcely noticeable by the astronomer. Such a star occasionally has an outburst, rendering it for a little time—sometimes for several years—as bright as Jupiter itself. One such case very luckily occurred within the spectroscope period. It was carefully examined by Huggins, and the result of the examination was to show that it was a star which had gone on cooling, or at all events had reached the lowest of its cooling stages, but suddenly became bright, because of an outburst of hydrogen. Bright lines broke out across its spectrum, showing that the incandescent gas which was in its atmosphere was at a higher temperature than at least the surface of the star itself. Now, this leads me to another and a curious remark about the lines of hydrogen which we see in the sun. Here is a portion of the solar spectrum as seen under particular conditions. It belongs to a solar spot, where of course the whole amount of radiation is less than that from the general body of the sun

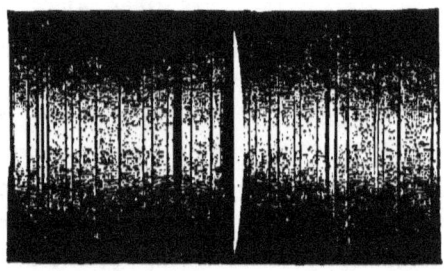

around it. Over that spot there must have floated an incandescent hydrogen cloud at a much higher temperature than the radiating portion of the sun at the spot, and therefore it was capable of radiating more of the hydrogen light than there was to absorb, so it behaved as a radiating medium instead of an absorbing one; and therefore the green line in the solar spectrum which is due to hydrogen came out as a bright line. After watching this phenomenon for a short time in this particular form, the observer saw it change into a line with a bright portion at one side and a relatively black portion at the other,—one part evidently due to radiation, the other to absorption, but both closely connected. Why did one half become bright and the other half black? The answer to that leads us to a study of a very curious kind, but I must defer this to another lecture. Meanwhile, as I have the electric apparatus at hand, there is another experiment I wish to show, though it is not directly connected with the subject I have been discussing.

I have here a cube of the well-known *Canary* glass, whose colour is due to oxide of Uranium. When I place it in the path of the rays from the electric arc it shows brilliantly its characteristic yellowish green light. But observe that this dark violet glass, when interposed between you and the cube, renders it practically invisible—in spite of its brilliant illumination. The violet glass is practically opaque to this yellowish green light. So far the experiment presents nothing very remarkable. But I now close the aperture of the electric lamp with the violet glass; and there, in the middle of the almost invisible beam which it allows to pass, is the cube of canary glass showing its characteristic colour almost as brightly as before.

Obviously the canary glass has *changed* the light which falls upon it:—for light can pass through the violet glass and afterwards develop the greenish colour to which the violet glass is almost opaque. This is one of the very beautiful experiments by which Stokes physically explained *Fluorescence* as a change produced by certain bodies on the refrangibility or, more directly, on the period of vibration of light.

Here is another exquisite experiment of the same kind. I illuminate (very feebly) a sheet of white paper by the radiation through the violet glass. With a brush dipped in a solution of sulphate of quinine, slightly acidulated by sulphuric acid, I write letters on the paper, and these at once shine out brilliantly with a light blue colour. This also is nearly invisible through the violet glass.

In both experiments the altered light is of lower refrangibility, *i.e.* of longer vibration-period, than the incident light—another instance of degradation of energy.

The point I shall first take up in next lecture is the point left unexplained to-day,—how it is possible for a line which was originally dark in the solar spectrum to broaden out and become bright, and then for one portion to become dark while the other portions remain bright.

LECTURE X.

SPECTRUM ANALYSIS.

Change of colour of Light by relative velocity of source and observer. Analogy from Sound. Causes of broadening of spectral lines. Spectrum of Solar Corona ; of Double Stars ; of Comets. Probable nature of Comets ; of Saturn's rings ; of the Zodiacal Light.

You remember I closed my last lecture by pointing out to you, for the second time, a diagram of a portion of the solar spectrum, in which we had side by side a bright line and a dark one, due to the same substance, namely, hydrogen. I told you that there is a very beautiful point of theory involved in the explanation of this phenomenon, and I proceed to give it. It generally goes by the name of Döppler's principle, but it depends upon precisely the same idea as that which led Römer to the discovery of the finite speed of light.

Let us take the simplest possible analogy. Suppose, for instance, that we had Mr. Perkins' steam-gun, and caused it to project bullets in the same direction, succeeding one another once every half-second. Then, if a target were held in the path of these bullets, it would of course be struck 120 times per minute. But suppose that the target were to move up towards the gun, while the gun still kept on discharging the bullets at the same rate, it is obvious that it would meet more bullets in the course of a minute than it would meet if it were standing still. If you were to withdraw the target gradually, keeping it always however in the line of fire, you would get fewer bullets per minute; and if you were to make it move away from the gun at exactly the rate at which the bullets are coming, then no bullets would reach it at all. One bullet would be in its neighbourhood, and would remain constantly at the same distance from it; for, in fact, the target and the bullet would be moving with the same rapidity.

Precisely the same thing may be observed in passing over a set of waves. If you were steaming through a set of waves in the direction in which the waves are going, it is quite conceivable that you may be steaming so fast as to be riding on the crest of a definite wave all the way; but steam a little more slowly, and you will see waves gradually passing you; steam still more slowly, and a greater number of them will pass you per minute. If, on the other hand, you are steaming so as to meet the waves, then you meet more than if you were not moving. The faster you go you meet the more waves per minute; and there is absolutely no limit to the number you may meet per minute, if you could only move fast enough to meet them. Now the impression, be it of pitch or of colour, that is produced upon the ear by sound, or upon the eye by a luminous radiation, depends entirely, so far as our present purpose is concerned, upon the number of these waves which meet them per second. Therefore, if we are moving towards a sounding body which is giving out a particular note, the number of waves which reach our ear per second will be greater than it would be if we were standing still, or (generally) if we were at rest relatively to the body. And as a higher note corresponds to a greater number of waves reaching our ear per second, it is obvious that in the former case, whether we are moving to the sounding body or the sounding body is moving to us, there will be a greater number of waves reaching our ears than if we were at relative rest ; so that we should perceive a higher pitched sound than what is actually given off by the sounding body. The experiment has been made by the help of a railway engine—first in Holland, and since in other countries—by stationing upon the engine a trumpeter, who had beside him a musician to control exactly the note that he should play. The musician, of course, was moving along with the trumpeter, and therefore heard precisely the note that the trumpet was sounding. The sound, however, was also heard by other musicians who were placed at the side of the line, and they noted that the faster the engine came up to them, the higher did they hear the note which was played by the trumpet ; and the faster the engine went away after passing them,—the faster it retreated from them,—the lower did this note appear to be. I have no doubt that you—at all events those of you who have paid any special attention to musical sounds—will be able at once to perceive this effect by means of such a simple instrument as this tuning-fork, even with such comparatively slight velocity as I can give it by swinging it in my hand. For the success of an experiment of this kind, it is better that you should close your eyes, in order that you may not associate the result with any movement which you may observe on my part ; and I shall endeavour to perform the experiment without making any noise which might indicate to you how I am moving, or whether I am moving, the apparatus at the instant. [Experiment shown.] You notice, then, that during the interval that I allowed the fork to sound, there was a period at which its pitch appeared to you to rise; then immediately afterwards it appeared to fall; then it rose again, and so on. We had a musical sound which was alternately higher and lower in pitch as I sharply moved the vibrating fork to or from you, and then, when the fork was held steady, we had the original sound. Now, precisely the same thing happens with regard to waves of light. If you move so as to meet more waves of light in a second, that will correspond to an impression upon your retina of a higher order of colour than if you were not moving to meet those waves, or if the body which was sending those waves to you were not moving towards you. Thus you see that the light which comes to us from a star is capable, not only, as I pointed out in my last lecture, of showing what chemical substances are incandescent in the atmosphere of the star, whether as giving out light on their own account or as absorbing portions from an otherwise continuous spectrum, but is also capable of pointing out to us whether the star is moving to us or from us; or still more minutely, whether a portion of its atmosphere is moving on the whole from us, and another portion on the whole to us. The first application of this by the spectroscope to the study of the relative motion of a star with reference to the solar system, was made by Mr. Huggins with reference to the dog-star. Of course, in order to find out from such experiments (which tell us only the relative velocity of the earth and the star in the direction of the line of sight) what the corresponding velocity of Sirius is with regard to the sun, it is necessary to consider in what part of its orbit the earth is during the observation, because when the earth lies in a line from the sun, making a right angle with the line drawn to Sirius, the earth is moving much faster or much slower towards Sirius than the sun is moving. On the other hand, when the earth is 180° from that position, it is moving slower or faster towards Sirius than the sun is moving. When the earth is so placed that Sirius and the sun are nearly on the same or on opposite sides of it, it is moving transversely to the line joining the sun and Sirius, and its motion relatively to the sun produces no modification of the observed phenomenon. We should have in such a case the full effect due to the relative motion of Sirius and the sun. Correcting, then, for the velocity of the earth relatively to the sun, Mr. Huggins found that the velocity of Sirius relatively to the sun is about twenty miles per second in a direction tending to increase their distance; so that ever since the time when Sirius was first observed, it has been steadily moving away from the solar system at the rate of something like twenty miles per second, and yet we have not the least documentary or other proof that the brightness or apparent magnitude of Sirius has become at all diminished in consequence. It has been leaving us at that tremendous rate, and yet so far is it, or has it been, from us all this time, that even this increment of distance, growing at such a tremendous rate, has made during historical periods no perceptible change in the amount of light that we receive from it.

The next application that was made of this principle was to verify the fact of the sun's rotation about its axis. It is obvious that, as the sun rotates about its axis in the same direction as the earth rotates, one portion of the solar equator, the portion to the left as we look at the sun in our northern hemisphere—the left-hand side of the sun—is coming towards us, and the right-hand side of the sun is going away from us. The sun's rotation about its axis takes place in what is called the positive direction; that is, the opposite direction to that of the hands of a watch, as looked at from the north pole side of the plane of the ecliptic. Now, although the sun's rotation is very slow,—that is to say, though the sun takes about twenty-six days to execute a whole revolution,—still, because of its enormous diameter, the linear velocity of all parts of its equator is very considerable: more than a mile per second. Therefore if we examine, by means of a spectroscope, the light which comes from, let us say, incandescent hydrogen at different parts of the solar equator, it should correspond to rather higher light (more refrangible rays—more waves per second) from the left-hand side of the sun's equator which is approaching us, than from the right-hand side, which is retiring from us; and, therefore, if we could by a proper optical combination place side by side, as coming through the same spectroscope slit, the light given out by incandescent hydrogen at these two extreme ends of the sun's equator as seen by us, then we should find of the two hydrogen lines, the one from the left-hand side shifted a little up in the scale, and the one from the right-hand shifted a little downwards. Therefore we should find, of course, the hydrogen line in different places of the two spectra; and by measuring the amount of displacement between the two, we could calculate what is the rate of the motion of these points in the sun's equator to us or from us, compared with the whole velocity of light in space.

Now, carry this just a step further, especially thinking

of the enormous velocities (which I discoursed upon in last lecture) with which these masses of flaming hydrogen are thrown out in explosions or eruptions from below the visible surface of the sun. Think of a rate of several hundred miles per second, or something like it, with which these masses of glowing gas are thrown out, and you can easily see that if something of the nature of, but incomparably superior in dimensions to, a cyclone, such as we have in our tropical regions, were taking place, accompanied by down-rushes of colder gas, and up-rushes of warmer gas, both of these being incandescent hydrogen, the general down-rush of the cold will correspond to absorption, and the up-rush of the hot to radiation. There will be cold gas absorbing, but going

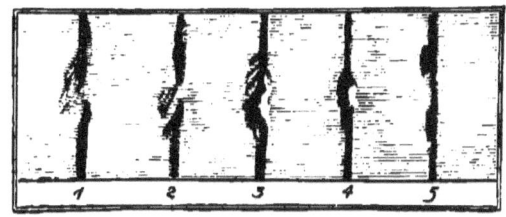

from us, and an up-rush of (on the whole) radiating gas which is coming towards us; and therefore we should find the absorption correspond to a lower position in the spectrum than the natural hydrogen line, while the bright line corresponding to the gas coming towards us will belong to a higher position in the spectrum; and so we account for the double line referred to in my last lecture, the lower half of it nearest the red being dark or due to absorption, and the other side being bright or due to radiation. Thus, even with a slit, the motion of these hydrogen clouds is easily seen by the blurred and broken form presented, whether by their absorption lines as seen on the spectrum of the solar surface; or their radiation lines as seen in the spectrum of the regions round the edge of the disc. Curious examples of these two phenomena are shown in the diagrams before you. Both represent appearances presented by the green line of hydrogen—in the first partly absorbent, partly radiating, the line is on the disc—in the second it is seen in a prominence, parts of which are moving with very great velocity. [Hence these pictures are *not* pictures of the prominence, as it would be seen by a telescope during a total eclipse, but pictures *distorted* by the Döppler principle.]

If we think for a moment of the whole light sent us by the sun, in which absorption by hydrogen far exceeds radiation by hydrogen, and think of the different relative rates of motion of different parts of the surface, we see a physical reason for broadening of the hydrogen lines altogether independent of pressure and cyclone currents. Hence a star in which the absorption bands are very broad may not necessarily have a dense atmosphere, but may be merely rotating rapidly about its axis. Thus caution is requisite in interpreting such appearances. And all the more so because Lord Rayleigh has called attention to the fact that even when a mass of incandescent gas is at rest as regards the spectator, its individual particles are in motion with sufficient relative rapidity to render a very narrow bright or dark line an impossibility. Even very rare hydrogen, if very hot, will therefore give broad absorption bands or bright lines. Other two causes, which may in certain cases lead to similar results, I must presently point out to you.

I may mention, before leaving this part of the subject, that Fox Talbot has proposed to apply the same principle to double stars, in order to find what is the distance of a physical system of two stars from us; at least when they have one common absorbing constituent in their atmospheres. If we can observe a double star, the plane of whose relative orbit passes (let us say, for example) nearly through the earth, then we may perform upon these two stars precisely the same operation as I have described with reference to the light coming from the two ends of the solar equator; and therefore of course we shall be able to tell what is the actual velocity of the one star in its orbit relatively to the other. We shall be able to calculate the relative velocity of the two, which is in fact the actual velocity of the one star in its orbit round the other; and knowing that actual velocity, we shall be able to calculate, from the observed periodic time, from the actual velocity thus determined, and from the apparent size of the orbit, not only what the actual size of the orbit is, but also how far that orbit is removed from us in order to appear so small as it does. So that by the help of this method, when properly applied, we shall be able to get perhaps a much closer approximation to the distance of various fixed stars from us than we can get by the only method hitherto employed, namely, by the determination of what is called their annual parallax. In fact, we may conceivably thus obtain a measure of the distance of stars so far off as to show no measurable, or even observable, annual parallax at all.

You see, then, that the light from a heavenly body can give us new information of very varied kinds,—information which was not sought nor even thought of as attainable until the introduction of spectrum analysis. We can find out, first of all, whether the light which it sends to us is light from a body of the nature of a solid or liquid, or at all events, a body of high general absorbing power, or whether it is light from a body of comparatively small and specific absorbing power, such as a glowing gas. Then, we can also tell—and this is perhaps one of the most curious of all the applications —if it be a glowing gas, at what pressure and at what temperature it exists in order to give off the spectrum that we find, because we can operate upon terrestrial hydrogen, etc., at various temperatures, and combine these with various pressures, and examine the spectrum under all such possible combinations, and then compare these variations in the spectrum with the varieties of hydrogen spectrum, which we get from the sun as a whole, from different parts of the sun's surface, and from various fixed stars. Therefore we are able to assign, not merely that it is this particular chemical substance, but also in what particular physical conditions it is found in order that it may give that particular kind of spectrum. Then we can tell, as we have just seen, the rate at which that particular radiating body is coming to us or going from us. The rate at which it is moving in a direction transverse to the line of sight is of course to be measured by ordinary astronomical processes, and therefore this fills up a *lacuna*—something that was wanting to ordinary astronomical processes, because we could tell perfectly well how a body moves transversely to the line of sight, but it is quite a novelty,—at all events when the body is one whose dimensions are invisible in the telescope,—to find the rate at which it is moving to or from us.

With reference to other possible causes (which are often at work—at least we may reasonably suppose so), besides variations of temperature and pressure, for the broadening of lines in the solar spectrum, let us think first of a particular effect that may take place in consequence of the currents of hydrogen gas in the sun's atmosphere. If part of the gas were going down slowly, part of it in a locality immediately contiguous going down faster, and then another stream going down still faster, then that part which was going down slowest would give the higher absorption line, and the part which is going down fastest from us will give the lowest absorption line ; and you would have, therefore, instead of the single definite narrow line which would be given by hydrogen remaining at rest, a broad band of absorption, parts of it corresponding to the different velocities of portions of the gas. All these absorption bands may fine off, as it were, continuously into one another ; so that although it is the same definite substance which is producing them all, it is producing them in different places in the spectrum, and filling with comparative darkness a definite breadth of the spectrum, because its different parts are moving from us with different velocities. That is another way in which the broadening of a band may occur in the solar spectrum.

But, as I said before, it may depend upon the fact that differences of temperature and pressure in general produce changes in the spectrum which a body gives. I shall come in another lecture to the consideration of the molecular theory of gases, when I shall speak of the particles flying about with very great velocity and impinging upon one another, and upon the sides of the containing vessel, and so producing what we call the pressure of the gas. Meanwhile, I shall anticipate so far as to say that when a gas is at the ordinary pressure of the atmosphere, each particle has to move a distance, let us say, of something like $\frac{1}{300000}$th or $\frac{1}{500000}$th of an inch on the average before it comes into collision with another particle, and is sent into a new path; but if you were partially to exhaust the gas in the receiver of an air-pump, there would be so much fewer particles in a given space that the length of the average path of any one particle, between one collision and the next, would be notably increased. On the other hand, if you were to compress the gas, then you might bring the particles so much closer together that no one would, on the average, be able to move more than, let us say, $\frac{1}{10000000}$th part of an inch, even at its very greatest excursion, before it would come into collision with another, and be sent into a new path altogether. And the more you compress the gas, the greater will of course be the number of such impacts for every particle in a given time, and therefore the shorter will be its average path between one collision and the next. Now the effect of heat also is to increase this number of impacts, because it makes the average velocity of the particles greater than before. The average square of the velocity of the particles corresponds in fact with what we call the energy of heat in the gas; and therefore corresponds nearly to what we call the temperature; so that as you compress the gas, you give its particles less way to go before they impinge upon one another, and as you still heat it under compression, you make them go faster and faster through the little range which they can compass before collision. Therefore, by these processes you make the collisions more numerous and more violent, and you also make the length of time during which a particle is in collision a larger percentage of the whole time of its motion. If it has only a collision now and then, it has a very small percentage of its time occupied by the collision, because the actual time of a collision is exceedingly short, and during the rest of the time it is moving free; but if collisions occur with very great frequency, then the time occupied in collisions becomes a serious fraction of the whole; and when a gas can be so far condensed as to approach the liquid state, its particles are scarcely ever free from collisions. Finally, when you get a body in the solid state, its particles are practically in a permanent state of collision with one another,—or, at all events, the time occupied in collisions is by far the greater part of the whole time. Now, during a collision, a particle of gas is not free; it is jammed against another or others; and therefore we may expect some modification to take place in the periods in which it is capable of vibrating. It is vibrating not by itself, but, as it were, only so far as the other or others will permit it, and thus the particles interfere with and modify one another's vibrations. Thus we see that if we have a very rare gas, we may expect that the spectrum which it gives off when heated will be in the main the spectrum due to the vibrations of the individual particles of the gas as they are flying about free from the others; but as we gradually compress it, the part of the whole time which is occupied in collisions increases, and then you do not get the pure spectrum of the gas,—what each particle would give on its own account, but in addition to that, you get the modification which is introduced by the action of one particle upon the next; and as you more and more compress it, and also as it is more and more heated, you get more and more of this interference of particles with one another. From free particles we get in general a few definite forms of vibration, corresponding each to a fine line in the spectrum, except in so far as this is modified by the relative velocities of the particles with regard to one another. When there are collisions, but not very numerous, we get slight modifications, generally as much in the way of increase of refrangibility as the opposite, so these lines broaden out on both sides. But as the amount of collision becomes more and more serious, and occupies more and more of the whole time, these effects spread themselves over larger and larger spaces in the spectrum; and so the effect of increased pressure and temperature is to make all the bands broader and broader, and finally, when we compress sufficiently, to reduce the gas to what is practically a solid, or at all events an incandescent liquid, the bands have so spread out that they have met one another, and you have in fact got a practically continuous spectrum. Thus the source of sunlight may not be a solid or even liquid globe—it may be merely a great thickness of very hot and highly compressed gas; in fact it seems quite possible that no portion of the body of the sun may be as yet even liquid.

Attending then to this, in addition to the other possible causes of modification which I have just mentioned, let us consider some of the data which are obtained by actual observation. I spoke to you in my last lecture about the spectrum of the incandescent part of the sun itself, and also of the protuberances which are seen during a total eclipse. But now let us consider the spectrum of what is called the corona,— the pearly white light which is seen round the body of the moon during a solar eclipse. There are parts of it, according to many drawings by accurate observers, which are obviously due to motes and ice crystals and various other things floating in the earth's atmosphere, because, of course, when you consider the enormous dimensions of the sun itself, it is quite certain that there can be no solar atmosphere (in the ordinary acceptation of the word), extending to a height of something like two or three diameters above his surface. Consider the enormous mass of the sun and its consequent attraction, and you will see at once that the idea of a solar atmosphere extending to anything like that distance is altogether preposterous. For in spite of the very high temperature at the sun's apparent surface, the density of the atmosphere there, due to the immense pressure, would in such a case be so great that a layer of moderate thickness from its lower part might easily have a density exceeding that of the sun as a whole; so that the sun would thus be in unstable equilibrium in a fluid denser than itself. Besides, there is a well-ascertained fact of quite a different character which goes against the notion altogether ; that is, that no two observers drawing such a corona, even at very short intervals of time from one another, or at very short intervals of distance from one another at the same time, ever draw at all nearly the same thing. That is a complete proof that at least the outer part of what has often been called the corona is a phenomenon due to the state of the terrestrial atmosphere in the observer's line of sight. But, even when the atmosphere is in its very clearest state, as it happily was in the south of India during the great eclipse of 1871, when most perfect observations were made, it is still found that there is a silvery light surrounding the sun, but extending to a height of, at the utmost, only fifteen or twenty minutes of arc above the dark circumference of the moon. That light has been analysed by the spectroscope, and its spectrum has been found to consist of two things,—one of them light from a glowing gas, the other reflected sunlight, so that the true corona owes its light to two sources. One is self-luminous gas, of whose composition I shall speak immediately ; the other, scattered particles which are capable of sending back sunlight. In fact, the spectrum of the corona as observed by Janssen, with an instrument specially contrived for the purpose,—a telescope with very large aperture as compared with its length, constructed for the special purpose of enormously increasing the brightness of the image of the phenomenon, —was simply a weak solar spectrum, not continuous, but having the dark lines, just like the spectrum of moonlight (which is merely reflected sunlight). But crossing it there were bright lines of hydrogen,—the C line, the F line, and the G line, which I described to you formerly [diagram, p. 192] : and, in addition to these, there was a green line, which cannot as yet be assigned to any known substance. That line appeared, in a detailed examination, to be given out even in the uppermost regions of the corona, regions farther from the sun than the highest in which hydrogen lines were seen. This would appear to indicate a gaseous element, one not only giving a simpler spectrum than hydrogen, but also a lighter element, capable of rising to higher elevations against the action of the sun's attraction. There must, then, be in the corona a solar atmosphere extending to a height of rather more than one-half the radius of the sun from his surface. It is possible it may extend still farther; but in addition to that, there must be matter which is capable of reflecting sunlight, and giving the continuous spectrum which Janssen observed.

Some very curious observations made in America upon the corona led to the detection of three bright lines which were found to coincide with lines which occur in the spectrum of the aurora. Now, it is a very singular fact that the terrestrial substance which gives these lines has not yet been discovered; and it is a problem of the most curious interest to us at present what substance it can be which,—incandescent by electricity no doubt,—during a terrestrial aurora, gives us the peculiar homogeneous green light which every aurora shows, and which is almost the only light given by the great majority of auroras. But the precise similarity and coincidence between the three auroral lines observed by one American observer, and the three lines observed by another American in the corona of the sun, seem to promise us wonderful information as to the similarity of the upper regions of the earth's atmosphere to those of the sun's atmosphere.

I shall now add a word or two to what I said in my last lecture with reference to double stars. I spoke to you about the spectra of fixed stars as indicating what may be called periods of life; but there are, besides, some very curious observations made specially upon double stars. All of you who have looked through even a moderately good telescope at double stars, must have noticed that many of such stars have extremely fine colours, very often directly complementary colours. Now, it was of course an interesting application of the spectroscope to find out to what these complementary colours are due. You can see at a glance when the spectra of the components of a double star are placed side by side, in what they differ. Now one of the first pairs examined showed for the first component the spectrum of a white star nearly; but the other component showed in its spectrum an enormous group of bands, cutting out almost the whole of the blue and green regions. Hence the group consists of a white star, with a practically red star revolving round it. But for an optical, or rather a physiological reason, of which it is not my business now to inquire the nature, a white body in the neighbourhood of a red body has a tendency to appear green. It is, then, merely an effect of contrast, as it were, that this double star appears in the telescope as an extremely fine green star, associated with an extremely fine red one. For when the spectroscope is appealed to, it tells us that there is a direct reason—obviously due to absorption in its atmosphere—for the one star's appearing red; but that there is absolutely no reason, except the physiological reason just alluded to, for the principal star's appearing green, for we see the spectrum it gives is almost devoid of absorption bands.

A few additional remarks remain to be made, chiefly with reference to comets. Unfortunately, the last very fine comet that was observed came before any one was prepared to apply the spectroscope to it; and, since spectroscopes have been in every observatory, no comets have appeared, except small and usually mere telescopic ones. There is no doubt, however, that the next fine comet[1] that appears will, especially by the help of spectroscopes, give us an amount of information as to the nature of comets immensely exceeding all that we have already gathered during thousands of years.

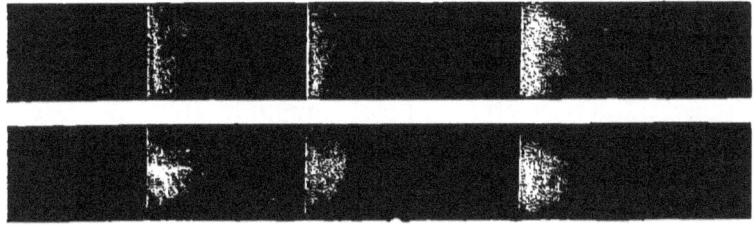

But such small comets as have been observed have given spectra which are extremely well worth noticing. Observations of these seem to show, first of all, that the tail of a comet gives a spectrum like that of the moon or other body illuminated by sunlight; in other words, that the tail of the comet is not self-luminous,—that it shines by scattered sunlight. But the head of the comet shows in general a spectrum which indicates the presence of glowing gas; that is to say, its spectrum is not continuous, nor is it visibly intersected by dark lines: it consists in general of a small number of bright lines

[1] These lectures were given in the spring of 1874, before the appearance of Coggia's comet. This was a magnificent object, but unfortunately ill situated for spectroscopic observation, having to be examined either very low in the horizon or in very strong twilight.

standing markedly out in relief from a feeble continuous spectrum. There (in the lower figure) is one of these—the spectrum of what is called Winnecke's comet, from the discoverer. It consists of three bright bands of light, each sharply terminated towards the red end of the spectrum, and shading away upwards to the violet end. Now Mr. Huggins, who first observed this, was struck by the resemblance of this spectrum (as he saw it in the telescope) to a terrestrial spectrum which he had noted before; and going over his note-book, he found it closely resembled the delineation of the spectrum of a hydro-carbon such as olefiant gas, rendered incandescent by passing an electric discharge through it. He then adopted the method to which I have already several times referred, of sending light from the two sources simultaneously through the upper and lower parts of the same slit, so that the spectra of light from the two sources should be placed side by side, and subjected to precisely the same series of refractions. When that was done the result was as shown in the diagram. The upper figure is the spectrum of some hydro-carbon, as given by an electric spark through the olefiant gas; the lower is the spectrum of the comet. Now, just as we had concluded that there is hydrogen in the sun's spectrum from the coincidence of the bright lines of terrestrial hydrogen with dark lines in the solar spectrum, here is a similar telling coincidence. Here is the coincidence of the three bands: a coincidence perfectly exact so far as the enlargement by the spectroscope enabled Huggins to measure it, not only of the bright terminations of these bands, but also in the gradual shading-off of each of them.

Now, this is a most remarkable phenomenon. It at

once suggests the question—How does the hydro-carbon get into this incandescent state in the head of a comet? A word or two on that subject may be of considerable interest, but we must lead up to it gradually. A great astronomical discovery of modern times is, that meteorites, the so-called falling stars, especially those of August and November, as they are called, follow a perfectly definite track in space, and that this track is in each case the path of a known comet; so that:— whether, as Schiaparelli and others imagine, the meteorites are only a sort of attendants on the comet; or whether, as there is, I think, more reason to believe, the mass of meteorites forms the comet itself:—there is no doubt whatever that there is at least an intimate connection between the two. The path of the meteorites is the path of the comet. Well, let us consider a swarm of such meteorites (regarded each as a fragment of stone), like a shower, in fact, of Macadamised stones, or bricks, or even boulders:—what would be the appearances presented by such a cloud? It must in all cases be of enormous dimensions, because the earth takes two or three days and nights to pass through even the breadth of the stratum of the November meteors. Consider the rate at which the earth moves in its orbit, and you can see through what an enormous extent of space these masses are scattered. Now, if you think for a moment what would be the aspect of such a shower of stones when illuminated by sunlight, you will see at once that, seen from a distance, it would be like a cloud of ordinary dust : and an easy mathematical investigation shows that it should give when sufficiently thick, except in extreme cases, a brightness equal to about half that of a solid slab of the same material similarly illuminated. The spectrum of its reflected or scattered light should be the spectrum of sunlight, only a great deal weaker. It is easy without calculation, but by simply looking at a cloud of dust on a chalky road in sunshine, to assure one's-self of the property just mentioned of such a cloud of dust or small particles. Remember that in cosmical questions we can speak of masses like bricks, or even paving-stones, as being mere dust of the solar system, and we may suppose them as far separated from one another, in proportion to their size, as the particles of ordinary dust are. Whether, then, it be common terrestrial dust, or cosmical dust, with particles of the size of brickbats or boulders, does not matter to the result of this calculation. Spread them about in a swarm or cloud, as sparsely as you please :—only make that cloud deep enough, and illuminate it by the sun, then it can send back one-half as much light as if it had been one continuous slab of the material. Now, look at the moon. You see there a continuous slab of material, and you know what a great amount of brightness that gives. And a shower of stones in space at the same distance from the sun as the moon, and of the same material as the moon, could, if it were only deep enough, however scattered its materials, shine with half the moon's brightness. Now, no comet's tail has ever been seen with brightness at all comparable to that of the moon ; and therefore it is perfectly possible, and, so far as our present means enable us to judge, it is extremely probable, that the tail of the comet is merely a shower of such stones, large or small.

But now we come to the question—How does the light from the head of the comet happen to contain portions obviously due to glowing gas, in addition to other portions giving apparently a faint continuous spectrum of sunlight, and perhaps also light from an incandescent solid ? The answer is to be found—at least so it appears to me—in the impacts of those various masses upon one another. Consider what would be the effect if a couple of masses of stone, or of lumps of native iron such as occasionally fall on the earth's surface from cosmical space, impinged upon each other even with ordinary terrestrial, not with planetary, velocities. In comparison with these latter, of course, the velocity of the shot of any of the big guns at Shoeburyness would be a mere trifle; yet we know that when a shot from one of them impinges upon an iron plate there is an enormous flash of light and heat, and splinters fly off in all directions. Now, mere *differences* among the cosmical velocities of the particles of a comet, due to different paths round the sun, or to mutual gravitation among the constituents of a cloud, may easily amount to 1400 feet per second, which is about the rate of a cannon-ball. Masses so impinging upon one another will produce several effects,—incandescence, melting, the development of glowing gas, the crushing of both bodies, and smashing them up into fragments or dust with a great variety of velocities of the several parts. Some parts of them may be set on moving very much faster than before ; others may be thrown out of the race altogether by having their motions suddenly checked, or may even be driven backwards ; so that this mode of looking at the subject will enable us to account for the jets of light which suddenly rush out from the head of a comet (on the whole, forwards), and appear gradually to be blown backwards, whereas in fact they are checked partly by impacts upon other particles, partly by the comet's attraction. Other very singular phenomena often presented by comets have recently been explained by a general rotation of the whole. And it is, of course, excessively improbable that a cosmical cluster of stones should not, whatever its origin, have a certain amount of moment of momentum in itself. Therefore, so far as can be said until we get a good comet to which to apply the spectroscope, this excessively simple hypothesis appears easily able to account for many even of the most perplexing of the observed phenomena. I must warn you, however, that this is not the hypothesis generally received by astronomers.[1]

There are various other phenomena in the solar system to which I might call your attention as capable of similar simple explanation, but I shall mention only two of them. The first is the wonderful appendage of Saturn, —what is known as Saturn's rings. There can be no doubt now that these rings are clouds of separate masses. This follows first from telescopic observation, which has shown us stars through one of the rings of Saturn, proving that there are numberless gaps in it, just as there are such gaps not only in the tail but in the head of a comet, through which we can see a star, even a small star, with almost absolutely undiminished brightness, and without refraction-change of apparent position. Again,

---

[1] [See *Proc. R.S.E.* 1868-9, and *Cosmical Astronomy*, V., *Good Words*, 1875. Recent researches, mainly due to Bredichin, have thrown very great additional light on this subject :—but have *not* added any new arguments in favour of the intrinsically improbable electrical hypothesis alluded to in the text. They have, however, made it possible that an action, somewhat akin to that which is shown by the *Radiometer*, may play a considerable part in causing the outrushes of tail-dust from the comet.—*Added to Third Edition.*]

mathematical calculation, founded on the laws of motion, has proved that rings like those of Saturn, if solid or liquid, would be broken up in a very short time by the enormous forces which are exerted upon them. The solid would either be broken up into pieces, or else it would as a whole go against Saturn on one side or another. The liquid would be broken up by enormous forced waves travelling round it, like the waves produced by a canal boat, which would go on increasing and increasing until they ruptured it. Clerk-Maxwell has shown, in his *Adams' Prize Essay*, that no hypothesis whatever will account for the form and permanence of these rings, except the supposition that they consist of clouds of stones, or fragments of matter of some kind or another, flying round, each almost like an independent member of a family of satellites, but still, of course, acting upon one another by their mutual gravitation. That mutual gravitation is, no doubt, sufficient to produce among them impacts with considerable relative velocity; so that it is possible that we may some day find bright lines in the spectrum of the light from the rings. Thus these rings of Saturn, like everything cosmical, must be gradually decaying, because in the course of their motion round the planet there must be continual impacts amongst the separate portions of the mass; and of two which impinge, one may be accelerated, but it will be accelerated at the expense of the other. The other falls out of the race, as it were, and is gradually drawn in towards the planet. The consequence is that, possibly not so much on account of the improvement of telescopes of late years, but perhaps simply in consequence of this gradual closing in of the whole system, a new ring of Saturn has been observed inside the two old ones,—what is called from its appearance the crape ring, which was narrow when first observed, but is gradually becoming broader. That is formed of the laggards, as it were, which have been thrown out of the race, and which are gradually falling in towards the planet's surface.

The second instance I refer to is the zodiacal light, which obviously cannot possibly be part of the gaseous atmosphere of the sun, nor can it be any solid or liquid body. It must be of the nature of detached portions of solid or liquid, floating as separate satellites, revolving about the sun, though by no means necessarily in orbits nearly circular. The spectrum of the zodiacal light has been examined. It is an extremely difficult thing to examine it; however, the task has been at least partially accomplished. The light is far too faint to enable even the most skilled observer, with the most perfect of our present instruments, to say whether there are dark lines across its spectrum or not. The spectrum has been found to be at least practically continuous; that is to say, it has been found to be probably that of reflected sunlight simply. Thus the zodiacal light reveals to us the existence of enormous amounts of small cosmical masses which have been somehow or other detached from comets or swarms of meteorites, and forced, whether by planetary attraction or by resistance, to revolve in orbits of moderate size about the sun. As they have been seized at different times and from different sources of supply, they probably move in all sorts of orbits—with all sorts of eccentricities and inclinations—somewhere about half of them probably going round in the opposite direction to that in which the planets move. Meteorites or aërolites, which every now and then reach the earth, may often be portions of this source of the zodiacal light. These scattered fragments, gradually resisted, or impinging upon one another, fall in age after age towards the sun's surface. They must thus form a supply, although an extremely small and inadequate supply, of potential energy, which has the effect of, to a certain extent, maintaining the sun's heat.

I must now take leave of this part of the subject, and I do so by recurring to what I said at the commencement of it. I began by saying that, after studying the laws of heat and thermo-dynamics, we should consider some very important cases of the transference of heat or energy from one body to another. We have already treated of the radiation of heat and the absorption of heat. Now we come to another case of the transference of energy:—the case in which energy is transferred continuously from one part of a body to another part of the same body; and here we must deal, first of all, with what is called conduction of heat. This subject was very fully worked out as a mathematical problem long before the period to which these lectures are professedly confined, but great additional information has been obtained about it within that period, and therefore I propose in my next lecture to give a brief sketch of the early development of it; and then to go more fully into the recent extensions and additions which it has obtained. Along with the conduction of heat I shall, virtually at least, treat of other things which, although having apparently no connection whatever with conduction of heat, really have precisely the same laws. These are the conduction of electricity, as, for instance, in a submarine cable, and the diffusion of a salt or an acid in a solution in water. Perfectly different as these phenomena appear to be, they are all, when treated mathematically, dependent upon the same differential equation (merely, of course, because their elementary laws, which are summed up with all their possible consequences in that equation, are of precisely similar form); and therefore by the change of a word or two, any statement made with regard to the one can be transformed into an equally true statement with regard to either of the others.

LECTURE XI.

CONDUCTION OF HEAT.

Fourier's Mathematical Theory. His Definition of Conducting Power. Analogy between Thermal and Electric Conductivities. Forbes's method and results. Ångström's method. Penetration of Surface temperature into the earth's crust. Analogy between conduction of heat and conduction of electricity. Diffusion also analogous to these. Diffusion of matter, of kinetic energy, and of momentum.

As I promised in my last lecture, I now proceed to a consideration of the subject of the conduction of heat.

A great deal was known about the conduction of heat before the period to which my lectures specially refer, but during that period a very great deal of quite unexpected information has been obtained on the subject. Perhaps it will conduce to the intelligibility of what I have to say about the new matter, if I briefly run over what was known about the time when Principal Forbes commenced his experimental inquiries into the question before us.

It was Fourier who first put the laws of conduction of heat into a perfectly definite mathematical form, and who invented, for the purpose of investigating detailed problems on the subject, a mathematical method of exquisite power. Fourier defined conductivity—the conducting power of a substance—in a manner which has not been improved since. He defines it, in fact, in this way. Suppose that you have a slab of unit thickness, but in surface practically infinite, composed of some material whose conductivity you wish to measure. Suppose one of its sides to be kept permanently at a temperature one degree hotter than the other side. Then, as we know that there is a constant flow of heat from a hot body to a colder one, there will be in this case (after things have settled down to a permanent condition) a definite rate of flow of heat through every unit of surface of the slab in a direction perpendicular to the slab. In fact, because we have supposed that the slab is of practically infinite extent, and that its surfaces are kept each throughout at a perfectly definite temperature, the flow of heat will necessarily be in the common perpendicular to the surfaces of the slab; and the measure of conductivity then, according to Fourier, is the number of units of heat which pass per square unit of surface of the slab from one side to the other in unit of time. You see, then, how all the different units come in. You have unit of length for the thickness of the slab: you consider the square of this unit—that is, unit of surface—as the portion of the slab through which the heat is passing. You have the unit of heat defined as the quantity of heat which can raise the temperature of a pound of water one degree. You have unit, that is one degree, difference of temperatures on the two sides of the slab, and you have unit of time during which the process of conduction is supposed to go on. Now, in an arrangement of the kind described, after a time, practically very short though theoretically infinite, the temperature will distribute itself permanently in this way :—The temperature will fall off steadily by a uniform gradient from the value on the one side to that on the other of the slab. It follows from this that the rate at which heat passes through the slab depends only upon two things,—the gradient or rate at which the temperature falls off per unit of length in the direction of its thickness, and the specific conductivity or conducting power of the material. Now, taking this datum, Fourier gave completely the mathematical formulæ which are necessary for applying it to any case—however complex—of the conduction of heat, in a solid of which the conductivity is not altered by temperature.

But this question very naturally arose—Is the conducting power of a substance the same at all temperatures? It had been assumed in Fourier's calculations that it was so; but Forbes seriously shook this assumption by pointing out a curiously complete analogy between the conducting powers of metals for electricity and their conducting powers for heat. It was found by experiment that those metals which conduct electricity well, also conduct heat well, and not only so :—Forbes pointed out that the order of conducting power for electricity is also, in the main, the order of conducting power for heat. [This observation of Forbes, which had been founded on the published experiments of other physicists, was confirmed by the experiments of Wiedemann and Franz, which were specially devised for the purpose of testing it.] Now, a point which has become of very serious importance of late years, especially in consequence of the development of submarine cables, is the very great change of electric conducting power of substances by change of temperature. Metals, in general, conduct electricity very much worse when hot than when cold; so that it occurred to Forbes that as there was an analogy—a *primâ facie* analogy, at all events

—between the conducting powers of different metals for heat and electricity, and as the conducting power for electricity is rendered very much worse by increase of temperature, so there might be an effect of this kind upon the conducting power of metals for heat. He therefore established a series of experiments, which, unfortunately, he lived to develop only as regarded the one metal, iron; but the results of these experiments were perfectly decisive in proving that the conducting power of iron for heat becomes worse and worse as it is hotter, and almost in the same proportion as it becomes by heat a worse conductor of electricity.[1]

I may say a word or two as to the process by which we investigate the conducting power, before I describe Forbes's experimental apparatus. Take an analogy first: suppose we consider the stock-in-trade of a certain business. There are two ways of investigating how that stock-in-trade may alter. One way consists simply in periodically taking stock, or going through the whole collection and seeing what it consists of. But there is another and equally good way, provided it could be carried out as well, and that is to keep an account of purchases and sales; so much has come in on the whole during the period; so much has gone out during the period; and the difference between the quantity which has come in and the quantity which

---

[1] [This, however, is true only of what is called the *Thermometric Conductivity*; in which the amount of heat conducted is measured in terms of the rise of temperature which it would produce in *unit volume* of the conducting substance at the temperature of conduction. But the specific heat in all substances alters with temperature. Thus Forbes's results are subject to serious modification when they are reduced to the usual thermal unit implied in Fourier's definition of conductivity.—*Note to Third Edition.*]

has gone out is the quantity by which the whole stock has changed during the period; so that there are these two ways of getting at it. Now, precisely the same idea is applied in ascertaining the conditions of the conduction of heat in a solid. We picture to ourselves a small portion in the interior of the solid, and for reasons of simplicity in calculation, we consider that small portion brick-shaped. We consider how much heat comes in through any one side, then how much during the same period of time goes out by the opposite side; and extend the process to the other two pairs of parallel sides. A mathematical expression can easily be formed for these various quantities, as I have already explained. They will be expressed in terms of the gradients of temperature, and the conducting powers (which may *not* be the same in all directions), parallel to the three sets of edges of the brick. But then there is the other way of looking at it. Instead of thinking what comes in and what goes out, think of how the temperature of the whole is altered during the period. You will see that in terms of the rise of temperature, the specific heat of the body, and the mass of the brick-shaped portion, we can make an independent calculation of how much heat has come in (of course on the assumption that no heat has been generated or destroyed within the brick). The latter of these expressions depends upon the rate of rise of temperature with time at any one point; the former depends upon the rates of increase of temperature per unit of length (or what may be called thermometric gradients) in three selected directions at right angles to one another. The gradients and the conductivity tell us how much comes in: the rate of change of gradient, per unit of length, and the conductivity, therefore, tell us how much more comes in than goes out; while the rate of rise of temperature, per unit of time, gives us another expression for the same quantity. It is the determination of relations between these two which is the object of every experimental inquiry on the subject.

Forbes's apparatus may be briefly described as follows:—These bars (*showing*), which were made for my own experiments, are made exactly of the dimen-

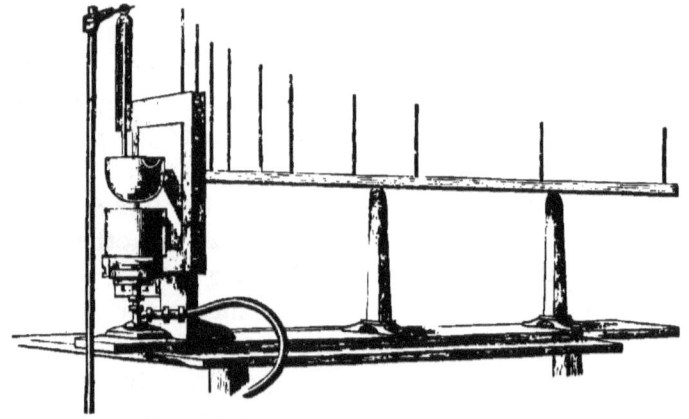

sions of Forbes's original bar. You will notice they are bars of 1¼ inch square section, and somewhere about 8 feet long, but that is not usually a matter of any great consequence. Along the length of each bar there are at intervals, first of three inches, and then of six inches, and finally of a foot, little holes cut vertically into the bar. In Forbes's iron bar these holes were simply filled with mercury, and the bulbs of thermometers were placed in them. In copper bars, and in German silver bars, such as those before you, it was necessary that these little holes should be lined with iron cups like arrow-heads, in order to prevent the mercury from attacking the substance of the bar. Now, matters being arranged in this way, a crucible was slid on, as you see, upon one end of the bar, and filled with melted metal, and a powerful lamp being applied to it, the temperature of the molten metal was kept as nearly as possible uniform for eight, or nine, or sometimes even ten hours. There was, therefore, a constant source of heat applied at one end of the bar, and all the rest of the bar was exposed simply to the air of the room. In the case of iron bars, Forbes found that even with the highest temperature to which he raised the crucible of molten metal, there was scarcely any perceptible rise of temperature in eight hours at the far end of the bar; but in my own experiments, I have found that because copper is so very much better a conductor than iron, it is absolutely necessary, if we keep the pot of metal at any moderately high temperature, to have a constant stream of cold water flowing over the farther end of the bar, in order to keep it from gradually increasing in temperature, even after eight hours' experimenting. However, the action of the cold water at the farther end introduces only a slight and simple modification of the formula, and in the mode of deducing the final results from it, but does not interfere with the mode of reasoning from the experiment.

The first effect of applying heat is to produce a gradual rise of temperature, which is of course observed first in the holes nearest the crucible. The thermometers farthest off are the last to give any indication of increase of temperature, and (after a steady state has been arrived at) are found to have risen the least.

What we wish to study now is the rate at which heat is being conveyed along; what our thermometers tell us is the temperature at different points of the bar. We must take care in making the deductions to remember that while our information is about temperatures, our conclusions require to be about heat.

Heat, then, gradually flows from the hot end of the bar to the cold one; and as the bar rises in temperature above the surrounding air, there is a loss of heat by radiation from its surface, and also by convection, by currents of heated air rising from the bar. This state of matters, strictly speaking, would go on indefinitely, approximating to a steady state. The steady state of temperature should (theoretically) never be actually arrived at; but practically in all our experimental work, a sufficient approximation to the steady state is arrived at in bars like these in at most eight or nine hours. After that time, provided we keep the temperature of the molten metal as nearly as possible steady, and provided the temperature of the air in the room remain unchanged, it is found that the thermometers have assumed definite readings from which they do not practically alter more than by very small fractions of a degree. There is then a steady state of temperature at every point of the bar, and that is the essence of the method. In such a steady state of temperature, of course, there is a steady thermometric gradient maintained at each point along the length of the bar; and it is found that practically we may assume, without risk of sensible error, the temperature to have the same value at all points of the same transverse section. The process I have just described to you may be applied to any thin transverse slice of the bar, so far as its supply, etc., of heat is concerned. First, in consequence of the greater steepness of the gradient of temperature at the warmer side of it, there is a greater quantity of heat passing into the slice by conduction than passes out of it by the same process. But because the temperature remains unchanged, that excess of heat must be lost by radiation and convection into the air. If, then, we could only measure how much heat is given off by radiation and convection from any given part of the bar, we should be able to measure how much more heat comes in than goes out in consequence of the difference of gradient at its ends. The temperatures are observed; from these the gradients and the difference of gradients can be calculated; multiply the difference of gradients by the conducting power, and by the area of the cross section, and you get the excess of the quantity of heat which goes in over the quantity which goes out per unit of time. Now that excess is precisely the loss from the external surface, also during unit of time. Forbes therefore made an addition to the usual experiment. He took a separate bar of the same material, of the same section, and in every respect similar to the first, only much shorter: and having heated this up, to a high uniform temperature, he allowed it to cool, simply noting its temperature after the lapse of successive equal intervals of time. Thence he calculated the rate at which it lost heat per unit of surface by radiation and convection together at each temperature. We have now by these two experiments an equation between two expressions,—one involving, besides known quantities, the conductivity which is unknown, the other consisting entirely of known quantities—and from this equation the conductivity is found. By that very ingenious

S

method, then, carried out by extremely careful experiments, the difficulty of which you may very well judge when I tell you that this pot of metal was usually heated to a temperature of somewhere about 300° or 350° C., and had to be kept sometimes for more than eight hours together without varying more than a single degree from that temperature, Forbes arrived at the conclusion which I have already stated, that the [thermometric] conducting power of iron falls off very rapidly with increase of temperature. He found that the conducting power at various temperatures is expressed by the following numbers, the units being the foot, minute, and degree Centigrade :—

0° C.,	0·0133
100° C.,	0·0107
200° C.,	0·0082

showing a remarkably steady diminution with increase of temperature. On looking at these numbers, we find that they almost exactly agree with the empirical law that the conducting power of iron for heat is inversely as the absolute temperature; that is to say, if you add 274° to each of these temperatures, you will find that the product of temperature so altered into the corresponding conductivity of iron is very nearly the same for each. Thus the conducting power is, as far as this determination allows us to judge, nearly inversely as the absolute temperature. This, if a general law, would appear to show that could we get an iron bar cooled down to a temperature of 274° under zero, its conducting power would become practically infinite; at least that, when the bar is almost deprived of heat, it has the power of conducting heat at an enormously great rate. That, of course, is arguing from results at a certain limited range of easily obtained temperatures to a range of temperatures on which we have not the least prospect of ever being able to make experiments at all.

I may mention, in passing, a curious form in which this semi-empirical statement as to thermal conductivity may be put. If we assume the principle of dissipation of energy to hold not merely for cases in which heat is altogether left to itself in a conducting body, but also in cases of artificially-sustained distribution of temperature, such as in this long bar of Forbes's, we have no difficulty in accounting for the fact that the conductivity may be inversely as the absolute temperature.

For (to take our earliest illustration of conduction) we conclude that any three consecutive slices of the infinite slab, of equal thickness, will have the least available energy when the absolute temperature of the middle one is the geometric mean of the temperatures of the others. Then the gradient will be as the absolute temperature, and (to make the flow of heat uniform) the conductivity must be inversely as the absolute temperature. This is on the assumption that the specific heat is unaltered by change of temperature, and must be modified accordingly.

I shall now say a word or two about a repetition of Forbes's experiments, and an extension of them to other bodies than iron, which has been carried on for some time in my own laboratory. You see there two copper bars, between which it would be exceedingly difficult for any of you, even with the aid of careful chemical analysis, to find much difference. The two bars are as alike as possible in their ordinary properties—in colour, specific gravity, elasticity, hardness, etc., and yet this mysterious energy, which we call heat, has far greater facility in passing along one of these bars than the other. One of them has somewhere about 40 per cent. greater conductivity than the other. Now, the only difference which we can detect between them is this,—that in the manufacture of one there seems to have been a very small quantity of oxide of copper mixed up with the molten mass, and this small trace (which it is difficult to measure by chemical processes) makes the metal a very much worse conductor of heat. These bars were obtained for the purpose of trying whether Forbes's analogy between different metals in their conducting powers for heat and electricity would extend to different specimens of the same metal. The bars were procured for me by Mr. Willoughby Smith, one being made of copper of very high, the other of copper of very low, electric conductivity. In fact that which conducts heat 40 per cent. better than the other conducts electricity about 73 per cent better.[1]

But then there comes in another and a very curious thing. You have seen that in all pure metals, as iron, copper, and so on, the electric conductivity falls off as the temperature rises. This is not the case with such an alloy as German silver. It is, in fact, used for electric resistance-coils because of the slight change produced in its electric conductivity even by serious changes of temperature. Here is a German silver bar of the same dimensions as the iron and copper bars. We find, on making the same experiments with it, that its conduc-

---

[1] [The experiments on the bars of copper and German silver, here described, had been *made* before these Lectures were delivered, but the extremely laborious process of deducing the conductivity from them had not been fully carried out. A full account of the results was given in *Trans. R.S.E.*, 1878.—*Note to Third Edition.*]

tivity for heat is much less affected by temperature than that of iron.

I have described one modern method by which conducting powers have been found. I have discoursed upon it so long that I must dismiss more briefly the other also modern method which has been applied to the purpose of experiment by Ångström, but which had been virtually employed in observations on a gigantic scale long previously to his time.

He, like Forbes, employs a bar, only instead of heating it steadily at one end, and waiting till a steady state of temperature has been set up in it, he produces a periodical change of temperature at one end. He heats it for a certain time, then cools it for an equal period, and repeats this operation until a steady state of *oscillation* of temperature has been practically attained at all points of the bar where observations are to be made. He observes at selected stations the range and the epoch of each wave of heat which thus travels along the bar, becoming less and less marked as it proceeds. This is in fact quite analogous to the process of telegraphing through a submarine cable. You apply one pole of a battery to the end of the submarine cable for a certain time :—then remove it—and so on :—and certain waves of electric potential run along the wire, by which intelligible signals are transmitted to the other end. Precisely the same thing, then, has Ångström done with reference to the conduction of heat by bars; and his method has given nearly the same conductivity as Forbes's for iron, which was the only metal experimented on by both. You will get some idea of Ångström's method and of the results deduced from it, if, instead of speaking of the more complex circumstances of the wave running along a bar, I speak of the simpler case in which we have a large slab of metal, heated periodically at one side, and kept cold at the other. Further, instead of metal, let us take the crust of the earth. Here is a diagram[1] prepared by Principal Forbes from continued observations of thermometers, whose bulbs were sunk, some in the porphyritic rock of the Calton Hill, within the Observatory grounds, some in the sandstone of Craigleith quarry, and some in the sandy soil of the Experimental Gardens. The curves on the diagram show the temperatures as indicated by these thermometers throughout the course of a whole year. The first thermometer at each locality has its bulb three feet below the surface of the ground; the second six feet below, the third twelve, and the fourth twenty-four feet under the surface. The observations are here figured in four groups, each containing three curves corresponding to the simultaneous indications at the different localities given by thermometers buried to equal depths under the surface. These thermometers (with the exception of one which was broken by the intense cold of the winter 1860-1) have been regularly read since they were buried. [This very valuable series of observations was interrupted by the wilful destruction of the instruments (September, 1876); but new ones have since been sunk, and the observations resumed.]

You will notice here that for the upper thermometer in the trap rock of the Calton Hill, you have the periodic wave of temperature lowest, not about the middle of winter, but about the middle of February. That is at

[1] It has not been judged necessary to reproduce this very elaborate diagram from *Trans. R.S.E.*, 1846, to which the reader is referred for fuller information on the question of terrestrial surface-temperature.

a depth of about three feet below the surface. We get the highest temperature at that depth about the middle of August; and so on down again to the lowest temperature in the middle of February next year. Now, another great point to be observed is that there is a considerable range of temperature at this depth; —for the lowest is somewhere about 39° F., and the highest somewhere about 54° F.; so that there you have a range of somewhere about 15° F. And remembering that the three lines which you see running alongside one another are for three such excessively different materials as porphyry, sandstone, and common light sandy soil, you see their general coincidence is very marked. They of course agree with one another in showing slight irregularities of temperature, due to periods of what we call 'change of weather' at the surface; but the ranges and epochs are not very widely different in spite of the variety of materials.

But see what a different state of things has been arrived at when you go only three feet farther down under the surface. There you find a far less range of temperature, though the mean temperature is nearly the same; the lowest temperature is now somewhere about 41°, and the highest somewhere about 51°, so that you have a range of only 10° altogether. When you go still farther down, to a depth of twelve feet, you will find a similar modification. [The irregularities here and there in some *one* of the three curves of each group, but not in the others, are evidently due to the percolation of water from the upper surface, or to some other purely local disturbance.] On the average, the twelve-foot observations show a range of from 44° to 49°, being a range of only 5°. And when you come down to the 24 feet thermometers you find barely a range of 1°·5 throughout the whole year.

Then remark, besides, that the minimum temperature arrives at the 6 feet thermometer somewhere about the beginning of March instead of the middle of February; it arrives at the 12 feet about the 20th of April; and at the lowest or 24 feet thermometer just about the middle of July,—that is to say, the winter's cold takes somewhere about half a year to penetrate twenty-four feet downwards into this kind of surface material.

Now this is almost precisely the same thing as—only on a much larger scale than—Ångström's experiment. The only difference is that Ångström had to allow for the loss of heat by radiation from the surface of his bar, while in the case I have been speaking of, there is no conduction except in a vertically downward or upward direction. Still you notice that the characteristics of the results are, on the whole, the same as those for the earth temperatures, that the ranges of the various thermometers diminish with great rapidity as you go farther and farther from the source of heat, and the periods at which the maximum and minimum arrive at any point are later and later as it is farther from the source.

Supposing the earth's crust to be of uniform material, and to have conducting power the same at all temperatures, the law made out by Fourier long ago was that as you go down successive depths in arithmetical progression, the range of the thermometers—for a simple harmonic wave of any period, such as for instance the annual one—should fall off in geometrical progression. If, for instance, at three feet down you had a range of 20°, and if at six feet down you had a range of only 10°,

CONDUCTION OF HEAT. 281

then on going down three feet farther, you should have a range of only 5°, and so on. Also the time at which you have what is called a particular phase of the wave of temperature (say its maximum or its minimum) should be later and later in proportion simply as you go farther down, so that if it be a month later at three feet, it should be two months later at six feet, four months later at twelve feet, and so on,—a month later for every three feet you go down. But notice that it would be so only on the supposition that the conducting power is the same at all temperatures.

In performing the bar experiment according to Ångström's method, the wave of temperature which passes the thermometers does not in general give, for its simple harmonic components, ranges diminishing in geometrical progression as we advance along the bar in arithmetical progression, nor are the periods of maximum constantly later and later by equal amounts for equal successive intervals along the bar; but it would be so if the surface-loss were very small, and the conductivity the same at all temperatures. Any such deviations then are due to these causes, and by them the amounts of the causes can be separately calculated.

Precisely the same statements that I made with reference to the distribution of temperature and the consequent flux of heat will apply, if instead of the word 'heat,' we use the word 'electricity,' and if instead of the word 'temperature,' we use the word 'potential,' which corresponds, in the theory of electricity, precisely to temperature in the theory of heat; so that when we write a mathematical formula to express the conduction of electricity in any body whatever, that formula will apply equally well to a corresponding case of the conduction of heat. There is no difference whatever between them till we come to the interpretations. We interpret a certain symbol in them to mean in the one case potential, in the other case temperature.

One of the most curious instances of imitation, on an exceedingly small scale, of what takes place on a very large scale, is suggested by this analogy. If I take a small piece of copper, an inch or so long, and, keeping one end of it in connection with a thermo-electric junction and a galvanometer, so as to measure very accurately any little changes of temperature that may arrive at that end, apply a lamp to the other end, just as you would apply to the near end of the Atlantic cable the pole of a galvanic battery; if I signal with this lamp just as the telegraph operator does with the galvanic battery through the Atlantic cable, exactly the same results may be produced on the galvanometers in the two cases, the tiny dimensions of the heat-conductor being necessitated by the time required to sensibly alter the temperature at the far end of the bar. You require to take a very short bar, indeed, in order to represent the phenomena on the same time scale, but you can have precisely the same effects in the two cases. And it is not at the ends merely, but at all similarly situated points in the two conductors, that the temperature and potential are proportional to one another at each instant of time.

Another illustration of the same thing is furnished by what you see here—the diffusion of a salt in solution through clear water. Take a tall cylindrical vessel, such as this, and having filled it with clean water, carefully boiled beforehand to exclude air from it, then, when the water is perfectly at rest, by means of a very long-necked funnel carefully pour in a solution of some deep-coloured salt, such as bichromate of potash, freely soluble in water, so that it lies as a horizontal stratum at the bottom, without disturbing more than slightly the water above it. We find that, in spite of gravity, some of the salt from the denser solution makes its way gradually up through the superincumbent column of water. But just as the rate at which heat passes from one part of a body to another depends upon the gradient of temperature, and just as the motion of electricity depends upon the gradient of the potential, so the rate at which this salt diffuses from one layer to another depends upon the gradient of strength of the solution. If there be any two places contiguous to one another where the strength is the same, then there would be just as much given out from the one side to the other as there is from the second side to the first, and therefore, on the whole, there would be no transfer; but where there is a denser solution placed immediately contiguous to a less dense one, then you have more given out per second from the dense solution to the rare one than you have from the rare one to the dense. And because this takes place according to precisely the same law as in the case of heat and electricity, any investigation whatever as to conduction of heat or electricity has a possible, or at least conceivable, application to some case of diffusion of a solid in a liquid which dissolves it. Now, in all these cases we have really been dealing with diffusion, whether of the particular kind of energy which we call heat, or of what we call electricity, or of matter itself, the law of the diffusion being precisely the same in the three cases. We call by the name conductivity, or conducting power, the numerical quantity which tells us how much heat or electricity passes in a given time across unit of area under a given gradient of temperature or potential ; so that, in the same way, we may speak of the diffusivity (if the word be permitted) of one substance in solution in another. This, again, is a definite numerical quantity, whose value is in every case to be determined by experiment. It depends not only upon the nature of the soluble salt, but also on the nature of the substance in which it is dissolved. The product of this diffusivity by the gradient of strength of the solution, gives you the quantity of salt which passes in unit of time from one side to another of a square unit in the liquid.

But it is not only matter and energy which can be so diffused ; we can also have diffusion of momentum. I conclude to-day with a brief discussion of that subject, because it will form an excellent introduction to the subject which I must take up in my next lecture. We will take an analogy, first published I believe by Professor Balfour Stewart, though it had occurred independently to others. Suppose there were a railway train passing through a station, and that a number of individuals who were waiting on the platform should jump into the train as it passed, and other passengers were to jump out of the train, what would be the effect on the train's motion ? Simply this—some parts of the train (because the passengers were at first virtually parts of the train) which had a certain amount of forward momentum, jumped out or left the train, taking their momentum with them. The train, therefore, as a whole, had less momentum, but also less mass, after these passengers left it than it had before. On the other hand, other passengers who had no momentum at all, simply stepped into the train as it passed them, so that the train gained no momentum by them, but gained mass. If a body gains mass, but does not gain momentum, it must be losing velocity. Thus the effect of those who jumped out is not either to increase or diminish the velocity of the train, unless they gave it a jerk on starting from it. If they jumped forward so as to give themselves more velocity than the train, they would retard the train to a certain extent, gaining in fact additional momentum, which is simply taken from that of the train. If they were to jump backwards, so as, if possible, to deprive themselves of any forward motion when they left the train, then they would quicken the train. But if we suppose them simply to step out transversely to the train, and take the consequences, they will leave the train with precisely the same velocity it had before, but it will have less momentum than before, because, although having the same velocity, it has less mass by the masses of the passengers who have gone out. But those who jumped in, if they jumped in in a direction transverse to the train, so as to have no momentum in the direction of the rails, must necessarily diminish the velocity of the train, because they cannot change its momentum, but they do change its mass, and if enough of passengers could jump into the train at once to enormously increase its mass, then they would be capable of reducing its velocity to any small amount you choose.

Carry the illustration a little further by supposing two long trains passing one another, and passengers stepping both ways between the two, it will be obvious to you that if this process could be continued long enough, it would end in destroying all difference of velocity between the two trains :—since *only* when their velocities are equal can there be no alteration of velocity by shifting a part transversely from one train to the other.

This illustration, of course, is a perfectly easily conceived one; but it will help us to understand what has until very lately been an extremely obscure subject,—viz., how it is that there can be such a thing as friction between portions of air; how it is that streams consisting each of detached particles flying about among themselves, can act as if they were solid bodies rubbing against one another. It enables us, I say, to explain upon what depends friction in fluids; in gaseous fluids at all events,—if it does not quite enable us to explain friction in liquids. Of course the explanation of friction in solids must depend on totally different principles. I shall, in my remaining lectures, explain the molecular motions due to heat in gases, and must therefore recur in part to our present inquiry.

LECTURE XII.

STRUCTURE OF MATTER.

Limits of Divisibility of Matter. In physics the terms great and small are merely *relative*. Various hypotheses as to structure of bodies—Hard Atom—Centres of Force—Continuous but Heterogeneous Structure—Vortex-atoms—[Digression on Vortex-Motion.] Lesage's Ultramundane Corpuscles. Proofs that matter has a grained structure. Approximation to its dimensions from the Dispersion of Light :—from the phenomena of Contact Electricity.

As I promised in my last lecture, I intend to take up to-day the consideration of recent results as to the ultimate nature and constitution of matter. This is a problem which has exercised the intellects of philosophers from the very oldest times. I have no doubt you are all acquainted with some of the speculations which, whether their own or not, Lucretius and others have given forth upon the subject. These, in many cases, possess even now some interest; but in comparatively modern times such inquiries have usually led to what may rather be called metaphysical than physical disquisitions. It became, in fact, the question of 'Yes' or 'No' for infinite ultimate divisibility of matter. Now that is a problem which, however simple it may appear to the metaphysicians, is at present quite as far from us physicists as it was in the time of Lucretius. We have made certain steps towards the knowledge of the nature of particles or molecules of matter; but as to the question of atoms,—that is to say, whether in going on dividing and dividing, if we could carry the process far enough, we should finally arrive at portions of matter which are incapable of further division,—that is a question, I say, whose solution seems to recede from our grasp as fast, at least, as we attempt to approach it.

There is a preliminary to all inquiries of this kind which, though obvious to every one worthy of the name of mathematician, is by no means obvious to an intellect (however naturally acute) which has no mathematical training. It is this :—that *there is no such thing as absolute size*, there is relative greatness and smallness—nothing more. To human beings things appear small which are just visible to the naked eye—very small when they require a powerful microscope to render them visible. The distance of a fixed star from us is enormous compared with that of the sun :—but *there is absolutely nothing* to show that even a portion of matter which in our most powerful microscopes appears as hopelessly minute as the most distant star appears in our telescopes, may not be as astoundingly complex in its structure as is that star itself, even if it far exceed our own sun in magnitude.

Nothing is more preposterously unscientific than to assert (as is constantly done by the quasi-scientific writers of the present day) that with the utmost strides attempted by science we should necessarily be sensibly nearer to a conception of the ultimate nature of matter. Only sheer ignorance could assert that there is any limit to the amount of information which human beings may in time acquire of the constitution of matter.

However far we may manage to go, there will still appear before us something further to be assailed. The small separate particles of a gas are each, no doubt, less complex in structure than the whole visible universe; but the comparison is a comparison of *two infinites*. Think of this and eschew popular science, whose dicta are pernicious just in proportion as they are the outcome of presumptuous ignorance.

I shall commence by briefly sketching one or two of the more plausible or justifiable opinions which have been enuntiated by various philosophers as to the so-called ultimate constitution of matter. The first—that which I have just referred to—is the notion of the perfectly hard atom. You meet with it, not only long before the time of Lucretius, but also in all subsequent time, even in the works of Newton himself. We find Newton speculating on this subject in his unsuccessful attempt to account for the extraordinary fact that the velocity of sound, as calculated by him by strictly accurate mathematical processes, was found to be something like one-ninth too little. We find Newton suggesting that possibly the particles of air may be little, hard, spherical bodies, which, at the ordinary pressure of the atmosphere, are at a distance from one another of somewhere about nine times the diameter of each; and, he says, sound then may be propagated instantaneously through these hard atoms, or particles of air, while it is propagated with the mathematically calculated velocity through the space between each pair. This is no doubt a very ingenious suggestion, and it enables him to get rid of the outstanding difficulty, because it virtually reduces the space through which the sound has to travel to $\frac{9}{10}$ths of what it other-

T

wise would have been; and therefore it enables him to add $\frac{1}{9}$th to the calculated velocity of sound. In fact, it virtually makes sound pass over $\frac{1}{9}$th more space in a given time than the mathematical investigation showed it should do. But, unfortunately for this explanation, it implies that sound should move faster in dense than in rare air at the same temperature. This is inconsistent with the results of direct measurement.

The true explanation, however, why the velocity of sound is different from that given by Newton's mathematical calculation, was furnished by Laplace, who showed that during the passage of a sound-wave through the air, the alternate compressions and dilatations take place so suddenly that the air has not time to part with the heat generated by the compression, or to supply the loss of temperature produced by the expansion, and therefore its pressure increases more when it is compressed, and diminishes more when it is dilated, than it would do if it were kept constantly at the same temperature. [In the language of Lecture V., above, the compressions and dilatations by which sound is propagated take place adiabatically.] And it is found by experiment that the amount of heat developed by compression of air, or the amount of heat absorbed in its expansion, is completely and exactly sufficient to account for the ninth which Newton found wanting; so that, although we have in this way Newton's authority for the supposition that there may exist ultimate hard particles through which sound or any motion may be transmitted instantaneously, the ground upon which he introduced them has now been found not to warrant that introduction, and therefore we are left as much in difficulty as before.

STRUCTURE OF MATTER.

There is one point, however, which should be noticed before leaving this speculation of Newton's:—that if the particles of matter be small hard atoms, whose size bears any finite ratio to the distances between contiguous ones, there must be a limit to the compressibility of any and every body. For instance, if Newton's suggestion had been correct; if the particles of air at the ordinary pressure of the atmosphere had a diameter equal to about $\frac{1}{9}$th of their mutual distance from one another (*i.e.* $\frac{1}{10}$th of the distance from centre to centre), then it is obvious that if you were to compress a mass of air into $\frac{1}{10}$th part each way in each of three dimensions, you would bring the particles in the various layers into contact with one another. That is to say, when a mass of air, taken at the ordinary pressure of the atmosphere, is compressed to $\frac{1}{10}$th of $\frac{1}{10}$th of $\frac{1}{10}$th, or $\frac{1}{1000}$th of its original bulk, then the particles are in contact with one another, like cannon-balls in a pile; and as they are, according to Newton's supposition, infinitely hard and incompressible, it would be impossible to compress the group further. Hence air could not be compressed to a smaller bulk than $\frac{1}{1000}$th of the bulk it has at ordinary pressures and temperatures.[1] Now, we know that air has been compressed by Natterer and others far beyond that, and therefore, tested from this point of view also, Newton's explanation or suggestion is seen to be quite

---

[1] This supposes the particles to be arranged in cubical order, so that each is in contact with only six others. But they can be packed closer, by arranging every three contiguous spheres in triangular order, so that each is in contact with twelve others. The density in this case is greater than in the former in the ratio of $\sqrt{2}$ to 1; so that Newton's group of spherical particles could be compressed to $\frac{1}{1414}$th part of the space they originally occupied. (*Proc. R. S. E.*, Feb. 1862.) This, however, does not invalidate the remarks above.

indefensible. It is obvious, however, that if there are such small infinitely hard particles as atoms, they must be in all bodies at a distance from one another, because, so far as experiment has guided us, there is no portion of matter whatever that is not capable of further compression by the application of sufficient pressure ; and of course, compression of a group of infinitely hard particles must presuppose that they have certain interstices between them, so that they are capable of being brought still nearer.

Another school of philosophers and experimenters, recoiling from the notion of the hard atom, took up the following idea. All that we know of atoms will be perfectly well accounted for if we dispense altogether with the notion of an atom—dispense with anything material in the ordinary sense of the word matter :—but suppose merely a centre of force, such as we are accustomed to in those mathematical fictions which we meet with in our text-books. Suppose, in place of an atom, a mere geometrical point, which can exert repulsive or attractive forces ; or rather suppose such forces tending towards or from a certain point, but nothing at the point; except, in some unexplained way, mass. So far as external bodies are concerned, this point will behave just as an atom would do. But here we are met by the gigantic difficulty of accounting for Inertia. This hypothesis was taken up and developed to a great extent by Boscovich, and was to a certain extent adopted in later times even by Faraday. It is, as I have stated at the outset, a mathematical fiction, but it is often an extremely convenient one for the purpose of enabling us to make certain mathematico-physical investigations of what takes place in the interior of bodies in their various states of solids, liquids, and gases.

Then there is a third notion—that the matter of any body, where it does not possess pores, like those, for instance, of a sponge (which obviously does not occupy the whole of the space which its outline fills), fills space, continuously, but with extraordinary heterogeneousness. But if, even in such a body as a sponge, we were to take a part so small as to be without pores, according to this notion, such a part is continuous but intensely heterogeneous.

In order to make this more plain, let us think of it on a greatly magnified scale; let us think of an enormous mass like a pyramid, for instance, or like the whole earth, built up of bricks and mortar; or, rather, like a huge mass of rubble, built of irregular stones, with mortar filling the interstices. Or, to take a body of intermediate size, suppose the moon were built up of materials of that kind. Looked at with our most powerful telescopes, it would appear to be perfectly homogeneous in texture. We could not possibly see the heterogeneity, the passing from the bricks to the mortar, or from the stones to the mortar, unless we could improve our telescopes so as to magnify enormously more than we can at present have any hope of. And unless we could obtain almost infinitely more perfect tranquillity of the air than has ever been noticed by observing astronomers (even when, as suggested by Newton, they have observed from the tops of high mountains), the possession of the requisite enormously high optical power would be impotent to reveal to us that the moon was not homogeneous throughout, but that it was really intensely heterogeneous when closely enough examined.

Now, what the moon presents to us at its distance of 240,000 miles—enormous from one point of view, but very small in comparison with other distances even in the solar system—very much the same thing is presented to us by a single drop of water. Our difficulty there is not on account of its distance, but on account of the extreme minuteness of the heterogeneity which we desire to measure. No microscope has yet enabled us to see anything of the nature of heterogeneity in a quantity of water, magnify it as we please, but yet there must be heterogeneity, and that of by no means inconceivably small dimensions, as you will see by many proofs which I shall soon proceed to give.

As, however, I am at present merely classifying the various more plausible suppositions as to the ultimate structure of matter, I do not now give that illustration, but leave it for a few minutes.

Finally, as the only other of these hypotheses which it is necessary to bring forward, I mention that quite recently suggested by Sir William Thomson—the notion that what we call matter may really be only the rotating portions of something which fills the whole of space; that is to say, vortex-motion of an everywhere present fluid.

The peculiar properties of vortex-motion were mathematically deduced, for the first time, by Helmholtz,[1] some fifteen years ago only, in a most beautiful investigation, in which he broke ground in a department of hydro-kinetics, the difficulties of which had, up to his time, kept mathematicians almost completely aloof.

So far, at all events, as concerns the motion of a perfect incompressible liquid, Helmholtz made out a great many thoroughly novel and excessively interesting pro-

[1] *Crelle*, 1857. Translated in *Phil. Mag.*, 1867.

positions, and upon these Sir William Thomson based his notion of possible vortex-atoms.

It will be necessary that I should give a brief sketch of the nature of these results of Helmholtz, in order that you may easily follow my explanation of Sir William Thomson's suggestion, and I do so the more readily because it is, or, at all events, it appears to myself to be, by far the most fruitful in consequences of all the suggestions that have hitherto been made as to the ultimate nature of matter. Especially does it give us a glimpse, at least, of an explanation of the extraordinary fact, that every atom of any one substance, wheresoever we find it, whether on the earth or in the sun, or in meteorites coming to us from cosmical spaces, or in the farthest distant stars or nebulae, possesses precisely the same physical properties. So convinced are we, by experiment and observation, that a particle of hydrogen, in the farthest nebulae, in the farthest stellar system, vibrates (when heated) in precisely the same fundamental modes, and in precisely the same periods, as it does in a Geissler's vacuum-tube in our laboratories; that, as we have already seen, any *apparent* exception to this is hailed as a certain source of information about the relative motion of such bodies with regard to the earth, and in some cases may give an invaluable method of obtaining their actual distances from us.

As a preliminary illustration, I shall show the formation of a simple circular vortex-ring: exhibiting one or two of its more important properties; and then we shall get rid of the apparatus for producing it as fast as we can.

The apparatus consists, as you see, of a very homely arrangement, merely a wooden box with a large round hole made in one end of it, while the opposite end has been removed, and its place supplied by a towel tightly stretched. In order to make the air which is to be expelled from this box visible, we charge it first with ammoniacal gas, by sprinkling the bottom of the box with strong solution of ammonia. A certain quantity of ammoniacal gas has now been introduced into it, and we shall develop in addition a quantity of muriatic acid gas. This is done by putting into the box a dish containing common salt, over which I pour sulphuric acid of commerce. These two gases combine, and form solid sal-ammoniac, so that anything visible which escapes from the box is simply particles of sal-

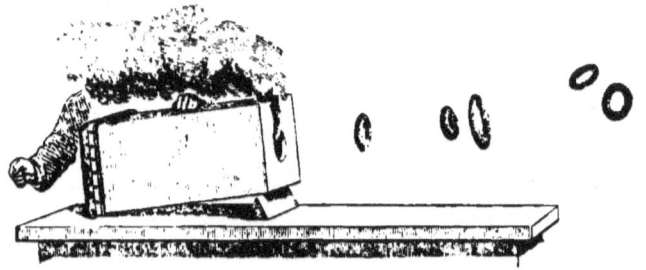

ammoniac, which are so very small that they remain suspended by fluid-friction, like smoke in the air. Now notice the effect of a sudden blow applied to the end of the box opposite the hole. There you see a circular vortex-ring, moving on its own account through the room as if it were an independent solid.

I shall now try to show you the effect of one vortex-ring upon another, just as I showed it here to Thomson, when he at once formed his theory. You notice that when two vortex-rings impinge upon one another, they behave like solid elastic rings. They vibrate vehemently after the shock, just as if they were solid rings of india-rubber. It is easy, as you now see (*shows*), to produce such vibration of a vortex-ring without any impact from another. All we have to do is to substitute an elliptical, or *even a square*, hole for the circular one we have hitherto employed. The circle is the equilibrium form of the simple vortex, and thus, if a simple vortex be produced of other than a circular form, it vibrates about that circular form as about a position of stable equilibrium. Another curious result which, as you see, is easily produced (*shows*), is to make one vortex-ring *pass through* another. Helmholtz showed theoretically that if two vortex-rings be moving with their centres in the same line, and their planes perpendicular to that line, then :— *first*—if they are moving in the same direction, the pursuer contracts and moves faster, while the pursued expands and moves slower, so that they alternately penetrate one another :—*second*—if they are equal and moving in opposite directions, both expand indefinitely and move slower and slower, never reaching one another. In fact, the one behaves to the other like its image in a plane mirror. And this, as you now see, is the fate of a vortex-ring which impinges directly on a plane solid surface.

Now, the first vortex-ring which you saw sailing up through the class-room, contained precisely that particular portion of air, mixed with sal-ammoniac powder, which had been sent out of the box by the blow. It was not merely sal-ammoniac powder which was going through the air, but a certain definite portion of the smoky air, if we so may call it, from the inside of that box, which, in virtue of the vortex-motion which it had, became, as it were, a different substance from the surrounding air, and moved through it very much like a solid body.

In fact, according to the result of Helmholtz's researches, if the air were a perfect fluid,—if there were no such thing as fluid-friction in air,—that vortex-ring would have gone on moving for ever.[1] Not only so, but the portion of the fluid which contained the smoke, which was, as it were, marked by the smoke, would remain precisely the same set of particles of the fluid as it moved through the rest; so that those which were thus marked by the smoke were, by the fact of their rotation, distinguished or differentiated from all the rest of the particles of air in the room, and could not by any process, except an act of creative power, be made to unite with the fluid in the room.

That is a point which appears to me to be one of the most striking characteristics in the foundation of this suggestion of vortex-atoms. Granted that you have a perfect fluid, you could not produce a vortex-ring in it; nor, if a vortex-ring were there, could you destroy it. No process at our command could enable us to do either, because in order to do it, fluid-friction is essentially requisite. Now, by the very definition of a perfect fluid, friction does not exist in it.

Thus, if we adopt Sir William Thomson's supposition that the universe is filled with something which we have no right to call ordinary matter (though it must possess inertia), but which we may call a perfect fluid; then, if any portions of it have vortex-motion communicated

---

[1] Of course, in air, fluid-friction, which depends upon diffusion, soon interferes with this state of things. But, in the experiment as shown, the ring (of some six or eight inches in diameter) was not sensibly altered by such causes in the first twenty feet of its path.

to them, they will remain for ever stamped with that vortex-motion; they cannot part with it; it will remain with them as a characteristic for ever, or at least until the creative act which produced it shall take it away again. *Thus this property of rotation may be the basis of all that to our senses appeals as matter.*

The properties which Helmholtz showed that such a vortex-filament must possess are these—first, that every part of the core of the filament is essentially rotating. A vortex-filament may have infinitely many shapes besides the simple one which I showed you just now. Unfortunately it appears impossible for us to form, even with an imperfect fluid like air or water, a vortex-filament of any more complex character than that simple circle. Theoretically, a vortex-filament can exist with any amount of knots and windings upon it, but then unfortunately we cannot devise an aperture by which to allow the smoky air to escape so as to give us such a knotted vortex. If we could devise the requisite form of aperture, and produce the vortex, then so long as the friction of the air did not seriously come into play, that vortex would retain its characteristics as completely as did the circular one which I sent out, and not only that, but it would possess that same elasticity about a definite form of equilibrium which you noticed the circular vortices possessed. They not merely keep their portion of air always in the vortex state, but they also have a definite form, in virtue of which they possess elasticity, so that when the form was altered for a moment by a sudden shock between two of them, each oscillated about its definite form, to which it finally settled down again.

In such a vortex-ring (as you will easily understand

by thinking how it came out of the round hole in the box), the motion of the particles of air is of this kind. Suppose it to be coming forward towards you, then every portion of the air on the inner side of the ring is moving forward, and every portion on the outer side is going backward, so that the whole is turning round and round its linear circular core. The air all about it is in motion according to a simple law which, however, I could not explain without mathematics—except in the particular case of that within the annulus, which is moving forward faster than the ring itself. I shall afford any of you who desire it (after the lecture)

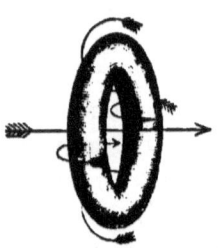

an opportunity of convincing yourselves of the fact. Each of you will find that, if he places his face in the path of one of these large air vortex-rings, there is no sensation whatever until the vortex-ring is almost close to him, and when it reaches him he feels a sudden blast of wind flowing through the centre of it. Thus, this vortex-ring not only involves in itself rotating elements which are thereby distinguished altogether from the other elements of the fluid, but it also is associated necessarily with other movements through the non-differentiated air, and especially a forward rapid current of air passing through its centre in the direction in which it is going.

Helmholtz showed that if vortex-filaments exist in a continuous medium, one of two things must follow:— either they must be ring-shaped—that is to say, endless —after any number whatever of knottings or twistings, the ends must come together; or else the ends must be in the free surface of the medium. Now, you can easily form half a vortex-ring. I dare say many of you have seen such a thing over and over again without thinking what it was. You must all have seen that when you draw a teaspoon along the surface of a cup of tea, and lift it up from the surface, there are a couple of little eddies or whirlpools going round in the tea rotating in opposite directions, the two moving forward (as do their sides which are nearest one another) in the direction in which the teaspoon was drawn. These two little eddies are simply the ends of a half vortex-ring. There can be ends in such a case as that, because these two ends are in the free surface of the liquid. A vortex-ring, then, cannot have ends, it must be a ring or a knot in fact, except these ends be in the free surface of the liquid; and if we adopt Thomson's notion of a perfect fluid filling infinite space, of course there can then be no ends. All vortex-rings—and therefore, according to Sir William Thomson, all atoms of matter—must necessarily be endless, that is to say, must have their ends finally united together after any number of convolutions or knottings.

Secondly, though this is really involved in what we have just seen, Helmholtz shows that such a ring is indivisible: you cannot cut it. Do what you like: bring the edge of the keenest knife up to it as rapidly as you please, it cannot be cut; it simply moves away from or wriggles round the knife; and, in this sense, it is literally an atom. It is a thing which cannot be cut: not that you cannot cut it; but that you cannot so much as get at it so as to try to cut it.

This idea of vortex-atoms enables us to explain a great many properties of matter; but it has introduced us unfortunately (perhaps I ought rather to say fortunately) to a series of mathematical difficulties of incomparably superior order to those suggested to us (at least at so early a stage) by any of the other ideas as to the nature of matter.

The fact is that Helmholtz's investigation, made fifteen years ago, was the very first attempt to take more than a single step into the wide and very difficult subject of hydro-kinetics, without the preliminary assumption that the motion should be differentially irrotational.

The subject of rotatory fluid motion is such a forbidding one, from a purely mathematical point of view, that no one had done more than take a look at it, as it were, until Helmholtz gave us these fundamental propositions;—splendid as they are, they are only a first step. Indeed, to investigate what takes place when one circular vortex-atom impinges upon another, and the whole motion is not symmetrical about an axis, is a task which may employ perhaps the lifetimes, for the next two or three generations, of the best mathematicians in Europe; unless, in the meantime, some mathematical method, enormously more powerful than anything we at present have, be devised for the purpose of solving this special problem. This is no doubt a very formidable difficulty, but it is the only one which seems for the moment to attach to the development of this extremely beautiful speculation; and it is the business of mathematicians to get over difficulties of that kind.

But there is more than this. In general, vortex-atoms, if they be at a moderate distance from one another, will not exhibit, in their behaviour to one another, anything of the nature of gravitation. That result at all events we can derive by our modern improvements of mathematics. How, then, is gravitation to be accounted for on this theory? The theory of vortex-atoms, being as it were complete in itself, must be rejected at once if it can be shown to be incapable of explaining this grand law of nature, which tells us that every particle or atom in the universe attracts every other with a force proportional to their masses conjointly, and to the square of their distance, inversely. Now the only even plausible explanation of gravity which has yet been propounded, was given long ago (at the beginning of the century) by Lesage of Geneva. He showed that gravitation can in all cases be accounted for by the not improbable supposition that, in addition to the gross particles of matter—I should, perhaps, rather say the particles of gross matter; but, as you will see, the term gross particles of matter also comes in as specially applicable to the hypothesis we are dealing with :—in addition to these grosser particles which are the atoms of tangible or sensible matter, infinite as these are in number, there is an infinitely greater number of much smaller ones darting about in all directions with enormously great velocities. Lesage showed that, if this were the case, the effects of their impacts upon the grosser particles or atoms of matter would be to make each two of these behave as if they attracted one another with a force following exactly the law of gravity. In fact, when two such particles are placed at a distance from one another, each, as it were, screens the other from a part of the shower which would otherwise batter upon it. If you had a single lone particle, it would be equally battered on all sides, but when you bring in a second particle, it, as it were, screens the first to a certain extent, in the line joining the two; and the first in turn screens the second, so that, on the whole, each of these two is battered more on the side opposite to the other one than it is on the side next the other one; and, therefore, on the whole, there is a tendency to bring the two together by the excess of battering outside over that inside. Now, it is a very easy mathematical deduction to show that the result of this is equivalent to an attraction, inversely as the square of the distance; and, therefore, that it exactly agrees with the law of gravity. But it is necessary also to suppose that *masses* of matter have a cage-like form, so that enormously more corpuscles pass through them than impinge upon them; else the gravitation action between two bodies would not be as the product of their masses.

This supplementary hypothesis requires, from Thomson's theory, an explanation of the supply of energy to these smaller particles; which must, of course, be smaller vortices. This has, as yet, not been fully given, though certain advances have been made. With a little further development the theory may perhaps be said to have passed its first trials at all events, and, being admitted as a possibility, left to time and the mathematicians to settle whether, really, it will account for everything already experimentally found. If it does so, and if it, in addition, enables us to predict other phenomena, which, in their turn, shall be found to be experimentally verified, it will have secured all the possible claim on our belief that any physical theory can ever have.

Although we cannot as yet attempt to settle the question whether there are atoms or not, we can, at all events, by the help of chemistry, take one step of immense consequence as regards the much simpler question of heterogeneity, to which I alluded a little ago.

We know that,—taking water as the most familiar of all substances to talk about,—we know that a drop of water can be divided and divided to an enormous extent,—whether to an infinite extent or not we don't know, but there is still a finite amount of division to which a drop of water can be subjected, in which the parts, which are finally separated from one another, are no longer the same as portions of the whole. By means of a galvanic battery, we can decompose water into its constituent gases. Now, this shows at once that there must be some limit to the division of a drop of water ; a point which we cannot pass without producing something different from water. The drop must be capable of finite, though excessively minute, division into parts quite alike, similar and equal to one another, but so small that if any one of these parts is again divided, the halves or parts of it could no longer be similar to one another. This similarity must necessarily go on to a certain extent, and to an extent far beyond what the microscope can show us ; but there must at last come a state of division when any further interference with one of the particles will make each fragment something other than water,—will take away a part of its oxygen or a part of its hydrogen, leaving too much of the one and too little of the other.

Whenever we come to that point, we have got down to what we may call the grained structure of the whole. Without any assumption at all about atoms, this is obviously something which exists, and which we reach in thought far sooner than we reach the atom. We know as yet nothing whatever as to whether oxygen and hydrogen consist of ultimate atoms or not; but we do know that water and all other compound substances certainly do consist of ultimate parts,—ultimate in this sense, that if you go any further with the division of them they cease to be parts of the substance.

Take a number of such similar parts of the substance as are capable, when mixed together in any numbers, of forming a mass of the substance. Perform the same sort of operation upon each of them, an operation which shall make it different from what it was before, by taking off a part of one of its constituents only; then if you mix all these similar parts together, you do not get the original substance. You get something else, if indeed you get a chemically stable compound at all, rather than a mere mixture of several different things. That shows, then, that without going infinitely far, you have arrived at a place at which heterogeneity begins to come in.

Now, it is a very great step indeed in science to determine at about what distance heterogeneity sets in. To take the former illustration which I gave, if the moon were built up, like a wall, of stones and mortar, into how many pieces must it be broken, so that these pieces shall not be, on the whole, quite similar in average materials to one another? A piece of massive brickwork of the size of this room could scarcely be distinguished from another piece of massive brickwork of the same size, if the bricks were laid in each in the same tactical order, and with the same thickness of intervening mortar. There might be a projecting fraction of a brick at one side more than at the other, and so on; but on the whole, a mass of brickwork of such a size—that is to say, a mass so large in its dimensions compared with an individual brick—might be talked of as homogeneous practically, or at all events as being simply and directly comparable with another piece of the same materials and size. But could we break down the mass into fragments :—after the mortar has become as strong, let us say, as the bricks :—fragments about the tenth part of the dimensions of the original bricks, then we should have pieces quite distinguishable; one, perhaps, wholly mortar, another wholly brick, and a third partly mortar and partly brick. There at last you have got heterogeneity, and marked heterogeneity, because two of the parts may have absolutely nothing in common, the one being entirely mortar, and the other entirely brick, although taken from the same solid mass, as nearly as possible from the same part of it, and of the same size.

It is quite obvious, then, that there is a most important investigation to be made here,—at what limit does this heterogeneity begin?—and also that it is already, to some extent at least, within our reach. The answer to that question has not yet of course been definitely given, but it has been approximately given by various authors, and on various grounds. Loschmidt was, I believe, the first who gave an estimate. Others have since given other estimates (very nearly agreeing with his) from the same point of view; and lately Sir William Thomson has not only given us an estimate from that point of view, but he has given us, besides, other three estimates from perfectly different grounds of reasoning, yet agreeing with one another, and with the slightly older results, as well at least as can be expected at the outset of so novel an inquiry.

As this is one of the most important of modern discoveries, I make no apology for dilating considerably upon these various arguments, the experiments upon which they are founded, and the results to which they lead us.

I may take first, then, a mode of proof depending on an investigation which, at the time when it was given, was hardly perhaps understood in its full weight. It was given considerably before the period to which my present expositions mainly refer. It was an investigation by the great French mathematician, Cauchy, of the motion of light in solid bodies and liquids; wherein he showed that, in order to account for what is called dispersion,—that is to say, the fundamental phenomenon discovered by Newton—the separation of the various colours of white light by refraction,—it was absolutely necessary, at least on his hypothesis, to take into account the effect of the distance between particles of matter (assuming that there are particles) upon the motion of the luminiferous medium. He showed that there could be no such separation of the various colours of light from one another, unless the distance from particle to particle of the substance through which the light was passing were comparable with the length of the wave of light, or at least were not infinitely small compared with it.

Thus, looked at from our modern point of view, Cauchy's result (*valeat quantum*) simply shows us that matter cannot be homogeneous. If matter were absolutely homogeneous, there might be refraction, but there would be no dispersion. All kinds of light would travel with the same velocity in glass, just as they did in the air outside ; and, therefore, the mere fact that the different kinds of light can be separated from one another in passing through a prism, gives, at least, a hint that the matter of the prism is heterogeneous, and that its heterogeneousness is not infinitely more fine grained than the length of a wave of any of the kinds of light which it enables us to separate in their courses.

Take this argument for what it is worth (especially if we connect with it the irrationality of dispersion), it gives us, at least, an approximation to the dimensions of the grained structure. For the average length of a wave of visible light is somewhere about, let us say, the 40,000th or 50,000th part of an inch. But the grained structure is probably very much more minute than the wave length. If it were not so, dispersion would, on Cauchy's hypothesis, be enormously greater than we find it. Again, we cannot suppose that it is much less than somewhere about the 10,000th part of the length of a wave of light ; that is,—in the course of one of these waves of light, which is only about the 40,000th part of an inch in length, there cannot be much more than 10,000 alternations from brick to mortar, as it were, and mortar to brick again :—and, therefore, by using the 10,000 and the 40,000 as factors, you may say that, in an inch, this heterogeneity, or the change, if you like to call it so, from a molecule to nothing, cannot occur much more frequently than somewhere about

400,000,000 times. 400,000,000 in the inch, then, will be a first very rough approximation to this heterogeneity or grained structure of matter.

The next which I take up, although not the next in order of time, is perhaps the next in simplicity of explanation to that of Cauchy. It is founded upon what is called the electricity of contact of different metals.

I shall first, by means of one of Thomson's electrometers, show you the fundamental experiment of this subject. I do so, because the experiment is one which, although perfectly well known in the time of Volta, has been steadily disbelieved since Volta's time, and is now received as true by a comparatively small number only even among physical philosophers. This is the electrometer, one of those instruments to which I alluded in my opening lecture as having been supplied in consequence of the practical demands of the time, and as having reacted in the way of enormously improving our means of making physical investigations. In order, first of all, to show that it is really acted upon by electricity, suppose I take this zinc plate, which is supported upon a glass stem, surrounded below by pieces of pumice moistened with strong sulphuric acid, and connected at present with the interior of the apparatus, and touch it with my handkerchief, you notice that the effect is to produce a deflection of the index-spot of light, which is sufficient to throw it off the scale. We know, however, that an electrified body put in contact with the ground loses all its free electricity. And you see now, that a touch of my finger on the zinc plate discharges its electricity, and the spot of light comes back to its original zero position. We shall study afterwards what kind of electricity was there developed. Meanwhile, I take a copper plate, which I hold by means of a glass handle, touch it first to remove any free electricity it may have, touch this zinc again to remove any trace of electricity, and then bring the two into contact. I take the copper plate from the zinc, and you notice the deflection of the spot of light to the left-hand side of the zero. I touch the copper, and apply it again. Taking it off, I have a still greater deflection; and I could go on doing this over and over again, and giving larger and larger quantities of electricity to the zinc plate. This is obviously the opposite kind of electricity to that which I produced in the zinc by touching it with my handkerchief. However, we shall presently test what kind of electricity it is. To show that this is a genuine charge of electricity, I touch the zinc, and you see the charge vanishes, and the electrometer reading is zero again. Let us vary the experiment, by putting the zinc plate where the copper was, and the copper where the zinc was. This time we are to observe the electrification of the copper plate when it has come in contact with the zinc. I make the contact, and withdraw it, and you notice now a deflection to the right-hand side instead of to the left. I perform it again, touching the zinc plate every time I have withdrawn it from the copper, and you notice the steadily-increasing deflection. We have thus established, as far as physical experiment can establish anything, that when a zinc and copper plate are brought into contact with one another, and then separated, the one is, in the usual language of the subject, charged with positive electricity, and the other with negative. To find out which is which, all that is necessary is to take a stick of sealing-

312 *STRUCTURE OF MATTER.*

wax, which produces what is called resinous or negative electricity when rubbed with a piece of flannel. If this be the same kind of electricity as that with which the copper plate remains charged, the effect of bringing the

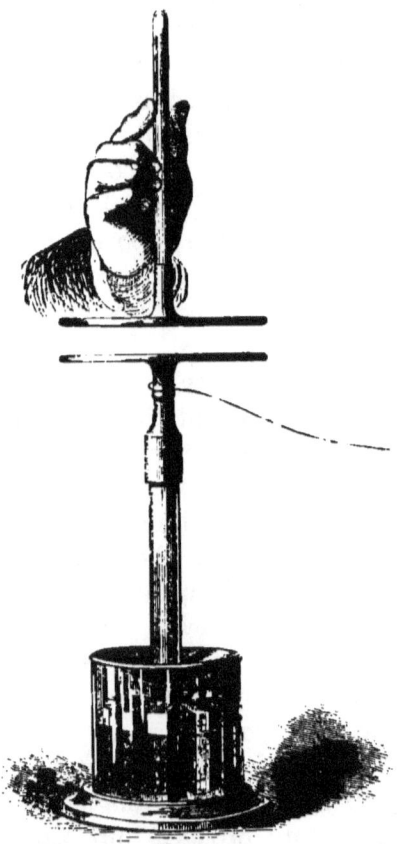

rubbed sealing-wax near the copper plate will be to repel that electricity into the electrometer, and therefore, of course, to increase the present deflection. On the other hand, if it should be the opposite kind of electricity, the wax will attract some of it out of the electrometer, and so the deviation of the spot of light will become less. I rub the wax and present it, and you notice we have a greater deflection ; therefore, copper is resinously or negatively electrified when it comes in contact with zinc, and zinc is, of course, positively or vitreously charged by the same operation. The large amount of the charge, in this experiment, depended upon the fact that the zinc and copper were put into metallic contact with one another over a large surface ; but I could have produced the same result in another way, more pertinent to my present inquiry. It would take up your time too much to exhibit the experiment, but if, in place of putting the plates in direct surface contact, I had supported one a small distance above the other, and brought a metallic wire from the one to the other, precisely the same effect would have been produced.

Now, to our application. When two bodies are electrified, one with positive and the other with negative electricity, they attract one another. Therefore, the mere fact of putting these metal plates in presence of one another, with a metallic wire, however thin, connecting them, introduces this force of attraction between the two, altogether independent of the force of gravity—a force of attraction due to the dissimilarity of the two metals. If they had been two plates of the same kind of metal, there would have been no such effect.

It was upon this experiment of Volta, which Sir William Thomson, by the help of his electrometer, was enabled to put beyond all cavil, that he founded another of those estimates of the dimensions of the heterogeneity of matter. The reasoning was of this kind.

There is a certain amount of attraction between the zinc and the copper plates when they have been put into metallic contact, however thin be the plates and the wire which is connecting the two. If we can measure the amount of electric attraction between the zinc and the copper plates, we shall be able to tell how much work would be done by this electric attraction if the zinc plate were allowed to come up to the copper plate. There is a certain force acting through a certain distance, and we can calculate how much work it would do. Suppose, then, that we take an enormous number of exceedingly thin plates of zinc and exceedingly thin plates of copper, and that we lay down first one plate of zinc, and then bring a plate of copper near it, but not touching it except by one corner, then there would be this electric attraction between the two. Allow the copper plate to fall down upon the zinc, and there would be a certain amount of work done. Then the upper surface would be copper. Bring a new zinc plate and repeat the experiment, and so on, you would get a pile of zinc and copper plates over one another. You can calculate how much work would be done in such a case, but it is easy to see that the quantity of work does not depend upon the thickness of the zinc or copper plates; so that, make the plates as thin as you please, the quantity of work done by the pile of zinc and copper plates, which together would rise to the thickness of an inch, will be greater and greater as there are more plates in it; therefore, make them thinner and thinner, and more numerous in the same proportion. You will get more and more work done by electrical forces on the same amount of mass, and, by carrying the process far enough, you would reach an amount of work sufficient,

if in the form of heat, to melt the whole of the zinc and
the copper. Now comes the application. It is obvious
from the reasoning I have just given that when zinc and
copper are put together in fine powder, it would be possible, provided there were no limit to their division, to
make the powders so fine that they would take fire, or
at least melt, on being mixed. Now we know by experiment the amount of heat which is generated when
copper and zinc are mixed intimately in the formation
of their alloy—brass—so that we can calculate how fine
must be their physical particles to account for their giving no more than the heat which is actually observed on
their thus coming together. That calculation depends on
a great many things, of which we have as yet no precise
measurement, so that such numbers as we can give must
be only approximate, but still we do not expect that they
are wrong by anything like 30 or 40 per cent., and that
is quite near enough for a first approximation in a question of such difficulty. It appears that if the thickness of the zinc and copper plates could be reduced to
about the 700,000,000th of an inch only, there would
be an amount of heat produced, by piling them together
alternately, which would more than correspond to the
quantity of heat which is produced by their chemical
action when they are melted together. Therefore we
see, from this way of treating the subject, that the
700,000,000th part of an inch is considerably under the
thickness to which you can reduce—if you could do it
—a plate of zinc or a plate of copper, without making
it cease to be zinc or copper as we know and handle
them. That is, in these metals the grained structure is
of dimensions considerably exceeding a 700,000,000th
part of an inch. That, you see, agrees perfectly well

with the 400,000,000th which we got from the other method, depending on Cauchy's work. But there are two other methods of making the approximation, which I shall contrast and compare with this one in my next lecture.

LECTURE XIII.

STRUCTURE OF MATTER.

Approximation to dimensions of grained structure from capillary phenomena —from properties of gases. Mathematical consequences of the supposition that a gas consists of constantly impinging particles—Gaseous Diffusion. Results of Maxwell's investigations. Physical reason of Dissipation —Andrews' results as to the continuity of the liquid and gaseous states of matter. Conclusion.

You will remember that in my last lecture I gave you in detail two of the methods by which Sir William Thomson approximated to the dimensions of the grained structure of matter. It remains, then, in commencing my present lecture, that I should briefly explain the other two methods.

You will remember that the first method was founded upon the dispersion of white light, or the separation of its various component colours by means of a prism. The second method was founded upon considerations of the amount of heat which would be generated by electrical action between particles of different materials when they were intimately combined together.

The third method is founded upon the forces employed in drawing out a film of liquid,—in fact (to take the simplest case), in blowing a soap-bubble. When we consider that the soap-film requires work to be done upon it in order to enlarge it—that is, to enlarge its surface,—and when also we consider that if we leave a portion of it open to the air, the contractile force of the film itself tends to make it shrink to smaller and smaller dimensions, and thus gradually to expel the contained air, we see that the film itself behaves to a certain extent like an elastic membrane, which requires work to be spent upon it in order to stretch it. But, just as a gas has no superior limit to its expansion, a soap-film has no inferior limit to its contraction.

Now it is a perfectly easy matter, if we know the tension of the film, to calculate what amount of work must be done in order to expand it from any given superficial area to any other given area; and by measuring the height to which the soapy water rises in a capillary tube of given bore, we can calculate what is the amount of surface-tension of the liquid. This is only one half of the amount of tension of a soap-film; for you must recollect that, thin as it is, it has *two* tensile surfaces with a layer of water between them. Hence, by experiments upon a capillary tube, we can tell what is this amount of surface-tension per linear inch. Then we can calculate from that what amount of work would be required to pull out a single drop of water until it was made into a film of any given thickness. The amount of work is, in fact, numerically the product of the tension per linear inch into the area of the surface.

Now it is found (in accordance with the fact that the surface-tension of water diminishes as the temperature rises) that in pulling out such a film, making it thinner and thinner, or spending work upon it against its molecular forces, it must become colder and colder, and that it would require a constant supply of heat in order to keep it at the temperature of the air. You will see then that there we have *data* from which it would be possible to calculate how much work would be required to pull out a drop of water into a film of a given thickness, keeping it always at a constant temperature.

This calculation has been made in terms of the thickness of the film, and it appears that if you pull it out to a thickness of the 500,000,000th part of an inch (supposing that could be done), you would require to spend upon it, besides the amount of work requisite to overcome the molecular forces, about one half as much energy in addition, in the form of heat, in order to keep its temperature from sinking; so that on the whole, including the heat which had to be communicated to it, the quantity of work spent upon it in the operation would be such that if it had all been applied to the drop of water in the form of heat, it would have been capable of raising it to a temperature of somewhere over 1100° C. Now, this amount of heat would of course wholly volatilise the water in an instant. It is therefore perfectly inconceivable that a film of water can be drawn out to such an excessive thinness without very great reduction of the molecular tension. But if the molecular tension is reduced, obviously we are coming to a state in which there are but a few molecules or particles in the thickness of the film, because as long as the film contains a great number of particles in its thickness, the whole tension of the film will remain sensibly unaltered.

Thus the only way of reconciling these two inconsistent things is by supposing that we have erred in assuming that, when we have made the film very thin, there still remains the original amount of molecular tension in it. Hence a film drawn out, if it were possible to draw it, to anything like the 100,000,000th part of an inch in thickness, cannot contain more than a very few particles of water in its thickness.

Then finally there comes the fourth argument, which is founded upon the behaviour of gases. I shall state merely the result just now, because I intend to devote the rest of my lecture this morning, or at least a great part of it, to the consideration of the motion of the molecules of gases. The result then that has been arrived at by several inquirers who have considered this molecular motion of gases is, that the average distance between the several particles of a gas at the ordinary temperature and pressure of the air must be something between the 6,000,000th part of an inch and the 10,000,000th part of an inch. This points (by a method which I shall presently indicate) to something rather greater than the 500,000,000th of an inch as the effective size of the particles.

Thus you see that all these various approximate estimates, differing no doubt considerably in the numbers which we obtain from them, still consistently point to something, not very largely differing from the 500,000,000th part of an inch, as being the distance between the centres of contiguous particles of matter in a liquid, or as being the measure of what I called in a former lecture the coarse-grainedness of what appears to our eyes, and even to our most powerful microscopes, to be absolutely uniform matter.

We have now got a notion of the dimensions of this coarse-grainedness, and it is possible also, by the properties of moving molecules constituting a gas, to find approximately the sizes of the individual molecules—not merely how far they are from one another, but how large is each particular molecule in comparison with the average distance between any two of them. This, as I shall presently explain, may be calculated from the theory of impact of elastic particles, or of particles repelling one another according to a high inverse power of their mutual distance.

I shall put the result, perhaps, in the most simple form by describing briefly the nature of the motion. Each particle describes a straight path with uniform velocity, but after going a certain distance it strikes another; then it goes off in a new direction, and after some time it strikes a third, and so on. But there is an average distance which it will pass through between every two successive collisions. Now, if we call that the mean free path, then the length of this mean free path, divided by the diameter of any one particle, has been shown by theory to be equal to the ratio of the whole space occupied by the gas to about eight and a half times the bulk of the whole particles.[1]

Now, looking at this, which is a mathematical result founded upon the supposition that the particles of a gas are little hard bodies constantly impinging upon one another, you will see that if we can get any mode of estimating what is the average distance which one of them must travel before it comes in contact with another, we shall know one of the quantities in the above proportion.

We also know another, the whole space occupied by the gas; and the whole bulk of the particles will depend upon their number and their diameters. These may be approximated to by supposing that in the liquid state the particles are nearly in contact. Thus an equation

---

[1] Clausius says 8, assuming that all the particles have the same velocity; but Maxwell, taking account of the law of distribution of velocities among the particles, gives $\sqrt{72}$, which is nearly 8·5.

can be formed, by which the diameter of a particle is given in terms of quantities which are all, at least approximately, known.

This calculation has actually been made, and the result is that the effective diameter of a particle must be something certainly not very different from one-250,000,000th part of an inch.

Then, of course, knowing the diameter of a particle, and the average distance between two contiguous particles, we can calculate how many particles there are in a cubic inch of any gas at the ordinary temperature and pressure.

Thus we can assert from measurement and calculation that the number of particles in a cubic inch of air in the ordinary state of the atmosphere is represented by a number which is approximately about 3×10^{20}. This number might have been written as 3 with 20 cyphers after it.

[It is a very much improved method of writing large numbers (when we cannot tell, or do not care, what are the figures in all the various places) simply to indicate the two or three most important figures of the number, and then to fill up with powers of 10. There is no use doing it in millions, billions, or anything of that sort :— in fact, billions, etc., have at least two quite distinct meanings with various sects of arithmeticians ; so we simply take the first two or three places of figures of a number, and indicate *how many places there are*: a far more excellent way.]

Be this as it may, 3 multiplied 20 times over by ten expresses a number not greatly falling short of the number of particles in a cubic inch of a gas at the ordinary temperature and pressure.

Now, you can see how it was that various scientific men were led to the conclusion which I mentioned in my first lecture as to the comparison between the actual size of the coarse-grainedness in a drop of water and the size of the drop itself.

If a drop of water is $\frac{1}{8}$th of an inch in diameter, and if those numbers I have just given represent the diameters of its individual particles, or its grained structure, what is the size of a body which bears to the whole earth the same ratio as one of those particles to this drop of water? The answer is that it must be something between about the size of a cherry or small plum and the size of a cricket-ball. Take on the average a good-sized plum or a small orange, then you get from that the approximation that as the large plum is to the whole earth, so is this coarse-grained particle to the drop of water; so that if we could magnify a drop of water to the apparent size of the whole earth as seen from the distance at which a single plum is just visible, we could just see its grained structure.

Now, what I have just said has led me to anticipate a little as to the molecular structure of gases. We have absolute proof that gases consist of particles of matter which are perfectly free and detached from one another, and which are constantly flying about in all directions. The best and simplest proof that we have of this is obtained by considering the process of mixture of one gas with another,—the way in which one gas diffuses through another; as, for instance, when any volatile substance in the form of vapour or gas is allowed to escape into a room, we find that it gradually mixes itself thoroughly with the air of the room. This diffusion takes place even if currents of air are prevented, so that at last there is almost uniform distribution of such a gas or vapour throughout the whole of any mass of air however great.

But, by means of a simple but extremely beautiful experiment, due to Dr. Graham, the late Master of the Mint, I can show in a most striking way the mobility and perfect independence (as it were) of the various particles of the same and of different gases. The apparatus consists of a glass tube, with a hollow ball of porous earthenware at its upper end, an arrangement very like an ordinary air thermometer, only that the bulb is not of glass but of porous clay. At present (*showing*) the whole apparatus is full of air, and there is air outside. You will notice that the open lower end being dipped under the surface of water, coloured for greater distinctness, the water stands as nearly as may be at the same level inside and outside, the reason of this of course being that there is precisely the same pressure in the gas inside as there is in the atmosphere outside. Now what is really going on here—what is keeping up this constant equality of pressure inside and outside? It is this: there is a constant stream of particles of air entering by all the pores of this porous clay, and a corresponding stream of particles of air coming out through them. We cannot, of course, mark individual particles of air, even could we see them, nor have we any test by which we could recognise a group of them. But to test the process which *must* be going on here, let us make a material difference between what is outside and what is inside the porous ball. Let us place this bulb with its air inside in an atmosphere of ordinary coal gas.

I shall easily obtain a quantity of coal gas by displacement in this beaker. (*Shows.*) I find that the gas is now escaping from beneath it, so that I know the beaker is full. Now watch the effect the moment that I surround this porous vessel containing air with the atmosphere of coal gas :—you see the bubbling which commences

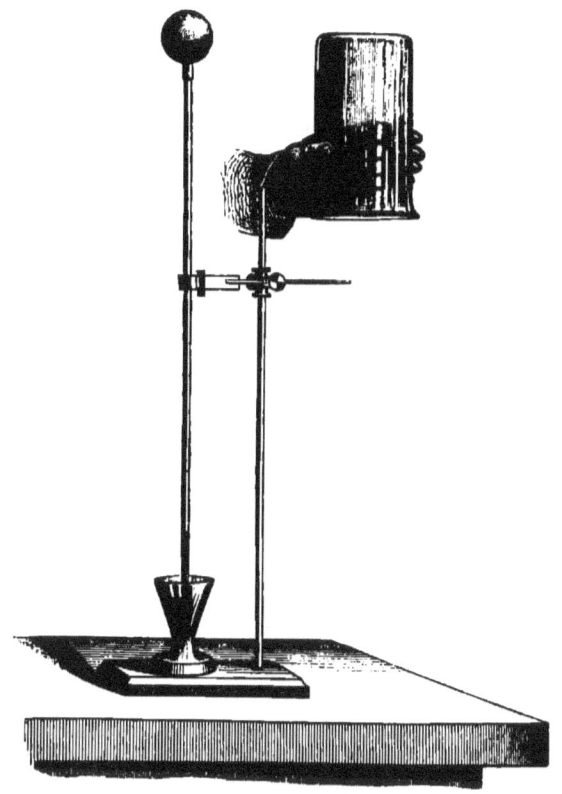

through the liquid almost instantaneously. Great quantities of gas are escaping from the liquid and from the open-mouthed bell. Next, if I remove the atmosphere of coal gas, you see almost instantly a rise of the liquid in the tube, so that the pressure in the inside has become at once notably less than the pressure outside. This process will go on for a considerable time, until we get as nearly as may be the restoration of equilibrium between the outside and inside pressures; but not to waste time over the experiment, although it is an extremely striking one, I shall simply, having a quantity of the coal gas still left in the beaker, put this atmosphere of coal gas once more outside the bulb. There! notice the instant increase of pressure inside; the coloured liquid falls at once in the stem. The moment I remove the beaker, there is an instantaneous diminution of pressure, and the coloured liquid rises.

This effect can be repeated as often as we choose, by simply putting on and withdrawing the vessel with the coal gas. Now it is perfectly obvious that the explanation of this experiment must depend on the fact that something gets in, in excess of what goes out, when I place the vessel full of coal gas outside the ball. There is then at once a great escape—a great out-rush of gas from the lower end of the tube, and that could only take place because of an increase of pressure; that is, an increase in the quantity of gas inside this vessel. In other words, the coal gas must be diffusing in through the porous clay at a greater rate than the air is diffusing out through it; and at how great a rate you can easily see, since the quantity which comes in through that comparatively small surface is sufficient to give us a rapid succession of large bubbles of gas passing out below. And you will remember here that it is only a differential effect that we are observing, because coal gas is constantly going in, but air is as constantly going out ; so that what we observe is merely the excess of the amount of coal gas going in over that of the air going out.

You can easily understand that methods can be devised for the purpose of measuring precisely the relative rates at which gases go in and come out through such a porous substance.

This is perhaps one of the most striking illustrations we can give of the great rate at which the particles of a gas must constantly be moving. It is also a complete demonstration of their comparative freedom from one another, except at instants when they come against each other in their motions.

Now, the notion that the particles of a gas are in rapid motion, and that it is by their impacts that gases press on other bodies, is a very old one. It was pointed out not long after Newton's time by D. Bernoulli; it was afterwards revived in this country by Herapath; but it is to Joule that we owe the first precise findings on the subject. Joule made a calculation as to what must be the velocity with which the particles of certain particular gases must move in order that, with the known mass of the gaseous contents of a vessel, the observed interior pressure on the vessel may be accounted for.

If we take a vessel containing a cubic foot and fill it with hydrogen gas at the ordinary pressure and temperature, the gas produces, by the constantly repeated impacts of its particles upon the walls of the vessel, a certain definite pressure which amounts to the ordinary barometrical pressure of 15 lbs. weight on the square inch. Now, we know what mass of hydrogen there is in the vessel. At what rate then must this rain, or rather hailstorm of particles be going on inside the vessel in order that these almost innumerable impacts may, when totted up for a definite time, amount to this definite observed pressure? That is a question, of course, of purely dynamical reasoning, and the result, as deduced by Joule, is certainly a very striking one. He arrived at this result, an absolutely certain result from his *data*, that the velocity of the particles of hydrogen must be about 6055 feet per second at 0° C. Of course you can see at once that this is a velocity far higher than that of a cannon-ball, or than that of any projectile which we can conveniently discharge; but in spite of the enormous velocity with which each individual particle of hydrogen is moving, there is such an enormous number of particles of hydrogen in a single cubic inch of space that no one particle can ever find anything like a free path from one side of a cubic inch of space to the other. It is certain to be met over and over again in its course by others.

Now, the number of such collisions which take place during a single second can be calculated from the rate at which one gas diffuses into another. If we take a vessel consisting of two large receivers full of gas, with a tube of known length and known diameter connecting them, and [opening a stop-cock in that tube] allow the two gases gradually to interpenetrate one another, we can, at the end of a measured time, close the stop-cock, and by a chemical process analyse the contents of each of the two vessels. Thus we determine how much of the one gas has passed into the other in a given time. We can repeat the experiment and allow it to go on for a different time, and so form a table of experimental results as to the rate of diffusion of one gas into another according to the time which has elapsed since the two were put into contact with one another. From such a table we can, by mathematical calculation, determine how far on an average any one particle of the one gas penetrates in amongst the particles of the other gas before suffering a collision.

Each particle advances a little way and then is driven aside or back, another gets possibly a little farther, and so on; but the average penetration can be calculated from the rate at which the gases are found to mix with one another, and therefore we can tell what is the average distance which a particle moves through between two successive collisions.

By applying calculations similar to those of Joule, but considerably extended by the use of more powerful mathematical methods, such as the methods of the theory of probabilities, Clausius first, and, a little later, but far more profoundly, Clerk-Maxwell, and still more recently Boltzmann, have arrived at very valuable results as to the motion of swarms of impinging particles. One of the results arrived at is that in a mass of hydrogen at ordinary temperature and pressure, every particle has on an average 17,700,000,000 collisions per second with other particles, that is to say 17,700,000,000 times in every second it has its course wholly changed. And yet the particles are moving at a rate of something like 70 miles per minute. So comes this curious problem— given that the direction of motion of a particle is arbitrarily changed 17,700,000,000 times in every second, and that the particle itself is moving 70 miles in a minute, where would it be at the end of a single minute, having started from any given place? In air the number of collisions for each particle is only about half as great as, and the average velocity only about one-fourth of, that in hydrogen.

You can see then to what exceedingly small quantities, or rather to what enormous numbers, because it is large numbers and not small quantities that we are really dealing with, such a question as this leads. I have already told you that in true physical science great and small are mere relative terms, so far as size or duration is concerned; but averages, such as we are now dealing with, are of no value except when the number of individual cases is very large.

The solution obtained is capable, as has been recently shown, of explaining almost everything that we know with reference to the behaviour of gases, and perhaps even of vapours. [The chief exception is in connection with the specific heats of gases. But the difficulty seems to have arisen from too hasty generalisations of the theory.]

As the direct results of Maxwell's investigation are very important, I may just point out the three which are of most importance. The first is this, that if you have a mixture of particles of different kinds, as you have for instance in air, particles of oxygen and nitrogen mixed with one another, then after they have gone on impinging on one another for a sufficient time to have attained the average state from which they will never afterwards much diverge, this will be the result—the average energy of motion of each particle is the same for each kind of gas; so that if one of these particles belongs to a light gas, that is to say, a gas whose particles are lighter or less massive than those of the other, it will, on the average, be moving with greater velocity; so that the energy of the motion of one particle of either gas is the same on the average as the energy of motion of one particle of the other gas. That is a result which can be obtained purely by ordinary mathematics of impact of elastic spheres, generalised so as to apply it in a statistical way to a great swarm of such spheres instead of a finite number. For the purpose of applying it by a statistical process to groups, it is necessary, in addition to the ordinary methods of such kinetic problems, to take into consideration the theory of probabilities. Thus we are led to classify the particles into groups, each group consisting of the particles whose velocity lies between given limits. The velocities proper to these groups range from zero to infinity, but the number belonging to such extreme groups is small compared with the number in the groups whose velocity is nearer to that which corresponds with the average energy. This average energy is the same in each of the two gases which are mixed.

Then, the second result which also follows from the theory of elastic particles impinging on one another is this, that if you have, as in the case of the air, oxygen and nitrogen mixed in any proportion whatever, the ultimate state of distribution which they will assume, after being mixed and having been left long enough to get to a nearly permanent state, is the same as if each gas, independently of the other, had distributed itself according to its own law of pressure and density in a vertical column. The oxygen and nitrogen of the air— so far, at all events, as gravity is concerned—are free to distribute themselves according to this law of mixture from the surface of the earth upwards; and, therefore, in whatever proportion you find them in a cubic inch near the surface of the earth, you would find (were it not for winds, etc., which tend to mix them) rather more Nitrogen and less Oxygen the higher you ascend.

Then another result bears upon the temperature of a vertical column of gas. Of course any particles which may be warmer than others, must be moving faster, because the temperature of a gas depends upon the rate at which its particles are moving. Now, Maxwell has shown that gravity has no tendency to make the quicker-moving particles go into any particular position, and the slower-moving ones into any particular position. In other words, an external force, such as gravity, does not tend to make the lower part of a column of gas any hotter than the upper part, or the upper part any hotter than the lower part. If you, for a moment, interfere with a state of things which has become nearly steady, or has arrived nearly at the average, as, for instance, by applying heat, then for a moment you make a portion of the gas physically lighter than another portion, and there, of course, gravity comes in, and the physically lighter part—bulk for bulk—ascends to the top of the column; but, if you leave the thing to itself, gravity has no tendency whatever to bring either the quicker-moving particles—that is, the warmer ones, or the slower-moving particles—that is, the colder ones—lower down or higher up; but, whether gravity acted or not, there will be the same average ratio of quick-moving particles to slow-moving particles in every cubic inch throughout the space. [And Carnot's reasoning, applied to this result, shows that as it is true for gases it must also be true for liquids and for solids:—so that gravity in no way *directly* interferes with distribution of temperature. *Indirectly* it does interfere, often in a marked manner, as in processes of convection.]

There is another extremely important point of this statistical question as to the particles of gases which I must carefully explain ; and it is this, how it happens that in the enormous bulk of the whole atmosphere of the earth these particles of oxygen and nitrogen, moving about amongst one another, should not by chance, at some place or other, operate on one another in such a way that in some particular cubic inch the particles of nitrogen might for a moment expel from it all the particles of oxygen, so that in virtue of the great extent of the earth's atmosphere, compared with the size of a particle of gas, there might be, at some definite instant, a region filled mainly with nitrogen, and other regions filled mainly with oxygen. Now the beauty of this statistical method is that it explains to us how such an event, though possible, never occurs. It is a thing which is in itself absolutely possible, but it never can occur in practice, because the probability of its occurrence is so exceedingly small. There is a probability (numerically measurable) for everything which is possible, but if that probability (reckoned in numbers) is as small as is the probability of the accident we are considering, we never expect to find it occur. And not only do we never expect to find it at any time, but we can say boldly from experience that it is never met with at all, however long our observations are conducted, or through however great an extent of space we conduct them. If you had originally in a box divided into two equal parts, nitrogen in the one part and oxygen in the other, and then allowed them to mix with one another, the probability that in any assigned time you would find all the nitrogen back again, even for a moment, in the space where it was originally, and all the oxygen back again in the space where it was originally, is certainly one which can be measured, but it is one which is so infinitesimally small that we know perfectly by experience that it never can be realised. The reason lies simply in the fact that there is such an enormous number of particles in every cubic inch. If there were only one or two particles of nitrogen and oxygen in each cubic inch of the atmosphere of this room, then it would be a thing not only realisable, but actually realised over and over again in the course of a single afternoon, that (suppose we could see these particles) we should every now and then see a space of a cubic inch, or it might be even of a cubic foot, occupied (for a single moment) wholly by particles of oxygen or by particles of nitrogen. It would be a different cubic foot or a different cubic inch the next instant, but still such spaces free from oxygen or free from nitrogen would be constantly occurring simply because there is a small number of particles in comparison with the whole space in which you are experimenting. But when you come to consider this number 3×10^{20} in every cubic inch, then you begin to see how it is that the number of particles is so enormously great, that if there were at any time the possibility of a small portion of space containing particles of oxygen only, or particles of nitrogen only, it could have a realisable probability in the case of so exceedingly small a fraction of a cubic inch only, that it would not be worth speaking of. You could not observe it. No doubt it does occur, but in an exceedingly small fraction, less than the millionth of the millionth of the millionth of a cubic inch. There may occur even in this room such portions wholly occupied by particles of nitrogen or by particles of oxygen for a moment, but only for a moment, in the sense of some millionth of a millionth of a second, but anything more than that could never present a reasonable probability ; and the reason of this depends simply upon the enormous number of particles which you have to deal with : for, the greater the number of independent movements the greater is the probability of their conforming nearer and nearer to the average distribution of amounts of velocity as well as to the average distribution of kinds of particles.

The only other point that I have time to take up is the consideration of certain most remarkable experiments which have been made recently by Dr. Andrews in extension of others made long ago by Faraday, and by Cagniard de la Tour, with reference to the connection between the gaseous and liquid states of matter. We have just seen how the gaseous state is completely, or almost completely, accounted for by the supposition of independent particles which are absolutely free from one another except at the instants when they impinge upon one another. Liquids possess the property of diffusing among one another just as gases do, only the rate of diffusion of liquids is exceedingly small compared with the rate of diffusion of gases, and therefore we conceive that the liquid state may be represented by a set of particles which are free from one another to a certain extent, but not nearly so much free from one another as they are in the gaseous state ; so that the particles of a liquid are so much nearer together in comparison with their diameters, that the average distance through which a particle of liquid can travel before it comes into collision with another is exceedingly small, even compared with the same path in the case of a gas

Therefore in a liquid we see why the diffusion is slower, because a particle can only go a much shorter way before it has its direction of motion completely changed by impact upon another. Now, Andrews has shown by experiment that it is possible to pass continuously —that is to say, without any apparent optical change— from a body which is obviously in a gaseous state to the same body obviously in a liquid state, and that is certainly one of the most remarkable experimental discoveries of modern times. Part of the phenomena had been obtained, but no complete experimental examination of them, by Cagniard de la Tour, many years ago. He had shown that if you take a quantity of ordinary ether in a strong glass tube, so as to fill about one-half the tube, leaving nothing but vapour of ether above it, and, sealing it up, apply heat to it, you may convert the whole liquid into vapour, so that the liquid contents wholly disappear; and this at a very moderate elevation of temperature. Cagniard de la Tour inferred from these experiments that sulphuric ether may be reduced completely to vapour by the combined action of heat and pressure in a volume a little more than double that of the liquid. The true explanation was first given by Andrews, who showed from his experiments on carbonic acid that when the liquid disappears as the tube is heated, some state of matter is produced which is certainly not liquid and certainly not ordinary vapour. Cagniard de la Tour performed the same experiment with water, but he found that for water it required so much stronger tubes and so much higher a temperature that, partly from the danger of explosions, and partly from the fact that water so heated attacked the glass chemically, it was next to impossible to make the experiment with precision.

STRUCTURE OF MATTER. 337

Andrews examined the subject in another way by actually compressing different gases, keeping them all the time at a definite temperature, and noting the relation between volume and pressure: then raising them to a little higher temperature, keeping them steadily at that temperature, and going through the same operation, and so on. In that way he was enabled, by direct experiments of the most beautiful kind, to lay down tables or charts which represented the relation between pressure and volume of these gases for any sets of temperatures between the limits through which he could experiment. The result of his inquiries may be easily seen if I roughly trace an approximation to some of his diagrams.

From what I told you of Watt's Diagram of Energy in a former lecture [*ante*, p. 110], and from what I have just said, you will see that Andrews made a detailed study of the *lines of equal temperature* for all the substances he experimented on. His first results were obtained from carbonic acid gas, and to them I shall recur:—but for ease of explanation, suppose we commence with vapour of water.

Take a cylinder containing a small quantity of superheated vapour of water at any temperature, and gradually compress it, keeping it at the same temperature, let us say, for instance, the temperature of boiling water. As we gradually compress, the volume diminishes and the pressure rises, the steam or vapour of water becomes more and more nearly saturated steam, until we come to a certain bulk which corresponds with its being all in the form of saturated steam. Then if we compress ever so little more, taking care to keep the temperature the same as before, we know what takes place:—the pressure does not alter, but some of the vapour is deposited in the form of water, and so we go on compressing without altering the pressure, but liquefying more and more of the contents, until we find the whole of the contents liquid. Then, still keeping the same temperature by proper appliances, we go on compressing; but we now find a very great difficulty in compressing any further, because we have liquid water to deal with, which now fills the whole of the part

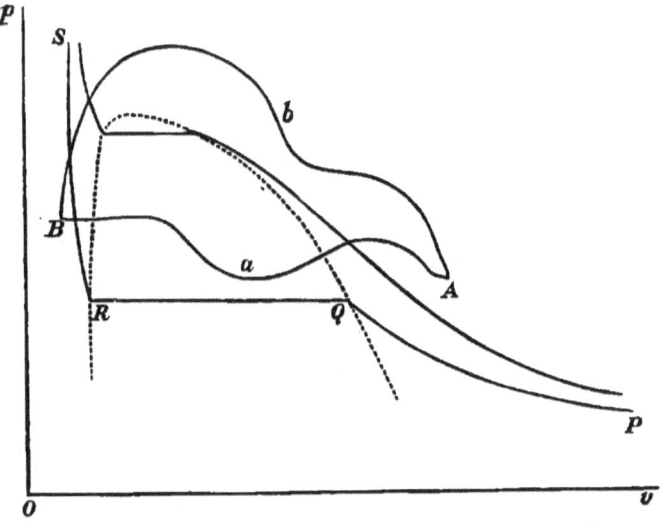

of our cylinder under the piston. The curve will now show very large increments of pressure for very small decrements of volume. Its course is *PQRS* in the diagram.

In the steam the pressure is small, while the bulk is large. It increases as we gradually compress, until we come to the perfectly definite pressure at which, at the given temperature, water begins to form. Then, however much more we may compress up to a certain limit, the pressure does not alter, so that our curve then runs along parallel to the horizontal line, and it does so until all the steam is converted into water; and then by further application of pressure we compress the water;—but exceedingly little—so that in reality our curve is now almost a straight line running vertically upwards.

That is the case for a temperature corresponding to the ordinary boiling point of water; but suppose we were to perform the same experiment on the same amount of steam at a higher temperature, we should find that we must now compress a good deal further before there is any deposition of vapour in the form of water, and that the water when formed occupies a somewhat larger volume, so that, on both accounts, the range through which we have part of our cylinder full of steam and part full of water, extends through a smaller horizontal space than before. Repeating the experiment at higher and higher temperatures, we come to smaller and smaller ranges of that kind, until finally we reach a temperature for water at which we cannot have part of the cylinder full of steam and part full of water.

Now you can see that if a curve, the dotted one in the figure, be drawn through all the points which terminate the horizontal straight parts of the various lines of equal temperature, we indicate on the diagram the region within which our given quantity of water can exist partly as vapour and partly as liquid.

Now we may, by proper application of heat and pressure, make the steam pass from one volume to another through any possible series of intermediate states. This series can, of course, be represented by a line on Watt's diagram. Take two points, A and B, both external to the dotted curve, but on opposite sides of it, so that in the condition denoted by A the water is obviously wholly in the form of vapour, at B wholly liquid. We may pass from one of these states to the other by any possible path, but for our present purpose only two need be considered, the first (AaB in the diagram) cutting the dotted line in two points, the second (AbB) passing wholly outside it. By the first course we have vapour alone till we enter the dotted curve—liquid alone after we pass out of it—liquid in presence of vapour while we are within it. On this no comment need be made—the passage from vapour to liquid is visible to the eye. But in the second course we pass from obvious vapour at A to obvious liquid at B, without its being possible at any point of the journey to say—the change is taking place.

You can begin it absolutely as gas, and bring it absolutely to the liquid state, but during the whole of the operation it is impossible to point out the instant at which it changes its character. And you can see perfectly well that the change can only be effected suddenly or in a manner marked to the eye, when you are performing the operations at such a temperature that the corresponding line in the diagram passes through the region in which the substance can exist in the liquid state in presence of its vapour. If you go beyond the limits at which this is possible, then you can cause the fluid to pass from the gaseous to the liquid, or from the liquid back to the gaseous state, with absolute optical continuity.

Or, to make it still more curious, suppose we go

through a complete cycle of operations from the state B to the state A, and back again to B. Begin, for example, with the lower of the two paths, BaA, and come back by the higher, AbB. We begin with water; then we have water and saturated steam about a, then superheated steam, till we reach A. On our way back we have no such stages—though when we have again reached B the contents are obviously water as at first.

Now the explanation of that phenomenon has not been fully given by the help of the dynamical theory of gases, but the mode of applying the dynamical theory to it has been, I think, successfully pointed out. One pregnant hint is given by the fact, noticed by Andrews, that the capillary surface of the liquid in contact with the vapour becomes less and less curved as the temperature rises to the *critical point* (*i.e.* the temperature at and above which the presence of liquid and vapour together becomes impossible), the curvature becoming *nil* when that point is reached. In all probability we want only a little further application of our statistical process to explain even this, which is perhaps the most curious fact that any investigation has ever told us as to the connection between liquids and gases.

Before concluding, I would just mention that there are a great many subjects to which I should have liked to direct your attention had time permitted. My apology for not having introduced them is partly want of time, partly the wish expressed to me that certain questions of history and priority should be fully treated. I have not been able to overtake during the time that I could devote to it nearly the whole of the programme that I at first put before me.

I shall just mention what some of these things were, and you will see that although I have taken up a great many interesting points, I have left still a great many others equally interesting. There is, for instance, the very interesting question of the explanation of vowel sounds and the qualities of musical notes; the whole in fact of Helmholtz's splendid investigations in Acoustics.

There is the whole subject of contact electricity, which I could only illustrate by a single experiment in my last lecture. There is the important subject, growing in importance every day, of Atmospheric electricity. There is Thermo-electricity, now almost a separate branch of physics:—Clerk-Maxwell's production of double refraction in viscous liquids; the connection between sun spots and terrestrial magnetism; the question of tides in the solid earth (whether the earth is plastic enough to have tides produced in its substance by the moon); we have the various proofs of the rotation of the earth; the connection between magnetism and light, as shown by Faraday's experiment and W. Thomson's and Clerk-Maxwell's investigations; the heating of bodies moving in what we call vacuum; the motion of light bodies produced by radiation; abnormality of dispersion, and so on; but even to merely enumerate all such would be, as it were, to double the length of this lecture. But such a statement shows, better than any comments, that we are dealing with a branch of science whose characteristic, as I told you at the outset, is persistent and ever more rapid extension.

LECTURE XIV.

FORCE.

Evening Lecture before the British Association, Glasgow,
September 8, 1876.

IT was not to be expected that I could, at short notice, produce a Lecture which should commend itself to the Association by its novelty or originality. But in Science there are things of greater value than even these—namely, definiteness and accuracy. In fact, without them there could not be any science except the very peculiar smattering which is usually (but I hope erroneously) called 'popular.' It is vain to expect that more than the elements of science can ever be made in the true sense of the word popular; but it is the people's right to demand of their teachers that the information given them shall be at least definite and accurate so far as it goes. And as I think that a teacher of science cannot do a greater wrong to his audience than to mystify or confuse them about fundamental principles, so I conceive that wherever there appears to be such confusion, it is the duty of a scientific man to endeavour by all means in his power to remove it. Recent criticisms of works in which I have had at least a share, have shown me that, even among the particularly well educated class who write for the higher literary and scientific Journals, there is wide-spread ignorance as to some of the most important elementary principles of Physics. I have therefore chosen, as the subject of my lecture to-night, a very elementary but much abused and misunderstood term, which meets us at every turn in our study of Natural Philosophy.

I may at once admit that I have nothing new to tell you, nothing which (had you all been properly taught, whether by books or by lectures) would not have been familiar to all of you. But if one has a right to judge of the general standard of popular scientific knowledge from the statements made in the average newspaper:—or even from those made in some of the most pretentious among so-called scientific lectures:—there can be but few people in this country who have an accurate knowledge of the proper scientific meaning of the little word FORCE.

We read constantly of the so-called 'Physical Forces' —Heat, Light, Electricity, etc.,—of the 'Correlation of the Physical Forces'—of the 'Persistence or Conservation of Force.' To an accurate man of science all this is simply error and confusion. And I have full confidence that the inherent vitality of truth will render the attempt to force such confusion upon the non-scientific public, quite as futile as the hopelessly ludicrous endeavour of the '*Times*' to make us spell the word Chemistry with a Y instead of an E. It is true that in matters such as this last a good deal depends (as Sam Weller said) 'on the taste and fancy of the speller:'—and sometimes even absolute error is of little or no consequence. But it is quite another thing when we deal with the fundamental terms of a science. He who has not exactly caught their meaning is pretty certain to pass from chronic mistakes to frequent blunders, and cannot possibly acquire a definite knowledge of the subject.

In popular language there is no particular objection to multiple meanings for the same word. The context usually shows exactly which of these is intended :—and their existence is one of the most fertile sources of really good puns, such as those of Coleridge,[1] Hood, Hook, or Barham. And there is no reason to object to such phrases as the '*force of habit*,' the '*force of example*,' the '*force of circumstances*,' or the '*force of public opinion*.' But when we read (as I did last week) in one newspaper that the 'force' of a projectile from the 81-ton gun has at last reached the extraordinary amount of 1450 feet : in another that the 'force' of a ball from the great Armstrong gun lately made for the Italian government is expected to average somewhere about 30,000 foot-tons :—and in a third that the water in the boiler of the Thunderer 'would in a second of time generate a " force " sufficient to raise 2,000 tons one foot high '—we see that there must be, somewhere at least, if not everywhere, a most reckless abuse of language. In fact we have come to what ought to be scientific statements, and *there* even the slightest degree of unnecessary vagueness is altogether intolerable.

Perhaps no scientific English word has been so much abused as the word 'force.' We hear of 'Accelerating Force,' 'Moving Force,' 'Centrifugal Force,' 'Living Force,' 'Projectile Force,' 'Centripetal Force,' and what not. Yet, as William Hopkins, the greatest of Cambridge teachers, used to tell us—'Force is Force,' *i.e.* there is but one idea denoted by the word, and all

[1] See *The Forked Tongue.*

Force is of one kind, whether it be due to gravity, magnetism, or electricity. This, alone, serves to give a preliminary hint that (as I shall presently endeavour to make clear to you) there is probably no such *thing* as force at all! That it is, in fact, merely a convenient expression for a certain 'rate.' If any one should imagine that 'three per cent.' is a sum of money, he will soon be grievously undeceived. 'Three per cent.' means nothing more or less than the vulgar fraction $\frac{3}{100}$. True, the '*Three Per Cents*' usually means something very substantial—but there the term is not a scientific one. Think for a moment how utterly any one of you, supposed altogether ignorant of shipping, would be puzzled by such a newspaper heading as '*The White Star Line*,' or '*The Red Jacket Clipper.*' No doubt some of our scientific terms approach as near to slang as do these:—but we are doing our best to get rid of them.

A good deal of the confusion about Force is due to Leibnitz, and to some of his associates and followers—who, whatever they may have been as mathematicians, were certainly grossly ignorant of some elementary parts of Dynamics—insomuch that Leibnitz himself is known to have considered the fundamental system of the *Principia* to be erroneous, and to have devised another and different system of his own. This fact is carefully kept back now-a-days, but it *is* a fact,[1] and (as I have just said) has had a great deal to do with the vagueness of the terms for *Force* and *Energy* in some modern languages. In fact, in their modern dress (with *Vis* everywhere rendered *Force*), the *Vis Viva*, *Vis Mortua*, and *Vis Acceleratrix* of that time have, in

---

[1] *Leibnitii Opera* (Dutens), vol. iii. 1768.

some of their Protean shapes, hooked themselves like entozoa into the great majority of our text-books.

Before dealing more definitely with the proper meaning of the word 'Force,' I must briefly consider how we become acquainted with the physical world, and how, consequently, it is more than probable that some of our most profound impressions, if uninformed, are completely erroneous and misleading.

In dealing with physical science it is absolutely necessary to keep well in view the all-important principle that

Nothing can be learned as to the physical world save by observation and experiment, or by mathematical deductions from data so obtained.

On such a text volumes might be written; but they are unnecessary, for the student of physical science feels at each successive stage of his progress more and more profound conviction of its truth. He must receive it, at starting, as the unanimous conclusion of all who have in a legitimate manner made true physical science the subject of their study; and, as he gradually gains knowledge by this—*the only*—method, he will see more and more clearly the absolute impotence of all so-called metaphysics, or *à priori* reasoning, to help him to a single step in advance.

Man has been left entirely to himself as regards the acquirement of physical knowledge. But he has been gifted with various *senses* (without which he could not even know that the physical world exists) and with *reason* to enable him to control and understand their indications.

Reason, unaided by the senses, is totally helpless in such matters. The indications given by the senses, unless interpreted by reason, are utterly unmeaning. But when reason and the senses work harmoniously together, they open to us an absolutely illimitable prospect of mysteries to be explored. This is the test of true science—there is no resting-place—each real advance discloses so much that is new and easily accessible, that the investigator has but scant time to co-ordinate and consolidate his knowledge before he has additional materials poured into his store.

To sight without reason, the universe appears to be filled with light—except, of course, in places surrounded by opaque bodies.

Reason, controlling the indications of sense, shows us that the sensation of sight is our own property; and that what we understand by brightness, etc., does not exist outside our minds. It shows us also that the sensation of colour is purely subjective, the only difference possible between different so-called rays of light outside the eye being merely in the extent, form, and rapidity of the vibrations of the luminiferous medium.

To hearing, without reason, the air of a busy town seems to be filled with sounds. Reason, interpreting the indications of sense, tells us that if we could see the particles of air we should observe among them (superposed upon their rapid motions among one another) simply a comparatively slow undulation of the nature of alternate compressions and dilatations. And our classification of sounds as to loudness, pitch, and quality, is merely the subjective correlative of what, in the air-particles, is objectively the amounts of compression, the rapidity of its alternations, and the greater or less complexity of the alternating motion.

A blow from a stick or a stone produces pain and a

bruise; but the motion of the stick or stone before it reaches the body is as different from the sensation produced by the blow as is the alternate compression and dilatation of the air from the sensation of sound, or the ethereal wave-motion from the sensation of light.

Hence to speak, as the great majority even of 'educated' people do, of what we ordinarily mean by light or sound, as existing outside ourselves, is as absurd as to speak of a swiftly-moving stick or stone as pain. But no inconvenience is occasioned if we announce the intention to use the terms light and sound for the objective phenomena, and to speak of their subjective effects as 'luminous impressions,' or 'noise,' as the case may be. In this case there is outside us energy of motion of every kind, but in the mind mere corresponding impressions of brightness and colour, noise or harmony, pain, etc. etc.

As another instance, it is obvious that we must be extremely cautious in our interpretation of the immediate evidence of our own senses as to heat.

Touch, in succession, various objects on the table. A paper-weight, especially if it be metallic, is usually cold to the touch; books, paper, and especially a woollen table-cover, comparatively warm. Test them, however, by means of a *thermometer*, not by the sense of touch, and in all probability you will find little or no difference in what we call their *temperatures*. In fact, any number of bodies of any kind shut up in an enclosure (within which there is no fire or other source of heat), all tend to acquire ultimately the same temperature. Why then do some feel cold, others warm, to the touch?

The reason is simply this—the sense of touch does

not inform us directly of temperature, but of *the rate at which our finger gains or loses heat*. As a rule, bodies in a room are colder than the hand, and heat always tends to pass from a warmer to a colder body. Of a number of bodies, all equally colder than the hand, that one will seem coldest to the touch which is able *most rapidly* to convey away heat from the hand. The question, therefore, is one of *conduction of heat*. And to assure ourselves that it is so, reverse the process; *let us*, in fact, *try an experiment*, though an exceedingly simple one; for the essence of experiment is to modify the circumstances of a physical phenomenon so as to increase its value as a test. Put the paper-weight, the books, and the woollen table-cloth into an oven, and raise them all to one and the same temperature, considerably above that of the hand. The woollen cloth will still be comparatively cool to the touch, while the metal paper-weight may be much too hot to hold. The order of these bodies, as to warm and cold in the popular sense, is in fact reversed; and this is so, because the hand is now *receiving* heat from all the various bodies experimented on, and it receives most rapidly from those bodies which, in their previous condition, were capable of abstracting heat most rapidly. However it may be in the moral world, in the physical universe the giving and taking powers of one and the same body are strictly correlative and equal.

Thus, the direct indications of sense are in general utterly misleading as to the relative temperatures of different bodies.

In a baker's oven, at temperatures far above the boiling point of water (on one occasion even 320° F.)—so high indeed that a beef-steak was cooked in thirteen minutes—Tillet in France, and Blagden and Chantrey in England, remained for nearly an hour in comparative comfort. But though their clothes gave them no great inconvenience, they could not hold a metallic pencil-case without being severely burned.

On the other hand, great care has to be taken to cover with hemp, or wool, or other badly conducting substance, every piece of metal which has to be handled in the intense cold to which an Arctic Expedition is subjected; for contact with very cold metal produces sores almost undistinguishable from burns, though due to a directly opposite cause. Both of these phenomena, however, ultimately depend on the comparative facility with which heat is conducted by metals.

Even, from the instance just given, you cannot fail to see that there is a profound distinction between heat and temperature. Heat, whatever it may be, is SOMETHING which can be transferred from one portion of matter to another; the consideration of temperatures is virtually that of the mere CONDITIONS which determine whether or not there shall be a transfer of heat, and in which direction the transfer is to take place. Bear this carefully in mind, because it has most important analogies to the results we meet with in considering the nature of Force.

It has been definitely established by modern science that *heat, though not material, has objective existence in as complete a sense as matter has.*

This may appear, at first sight, paradoxical; but we must remember that so-called paradoxes are merely facts as yet unexplained, and therefore still apparently inconsistent with others already understood in their full significance.

When we say that matter has objective existence, we mean that it is something which exists altogether independently of the senses and brain-processes by which alone we are informed of its presence. An exact or adequate conception of it, if it could be formed, would probably be something very different from any conception which our senses will ever enable us to form; but the object of all pure physical science is to endeavour to grasp more and more perfectly the nature and laws of the external world, using the imperfect means which are at our command—reason acting as interpreter as well as judge, while the senses are merely the witnesses, who may be more or less untrustworthy and incompetent, but are nevertheless of inconceivable value to us, because they are our only available ones.

Without further discussion we may state once for all, that our conviction of the objective reality of matter is based mainly upon the fact, *discovered solely by experiment*, that we cannot in the slightest degree alter its quantity. We cannot destroy, nor can we produce, even the smallest portion of matter. But reason requires us to be consistent in our logic; and thus, if we find anything else in the physical world whose quantity we cannot alter, we are bound to admit it to have objective reality as truly as matter has, however strongly our senses may predispose us against the concession. Heat therefore, as well as Light, Sound, Electric Currents, etc., though not forms of matter, must be looked upon as being as real as matter, simply because they have been found to be forms of energy—which in all its constant mutations satisfies the test which we adopt as conclusive of the reality of matter. We shall find that this test fails when applied to Force.

But you must again be most carefully warned to distinguish between heat and the mere sensation of warmth ; just as you distinguish between the motion of a cudgel and the pain produced by the blow. The one is the *thing* to be measured, the other is only the more or less imperfect reading or indication given by the instrument with which we attempt to measure it in terms of some one of its effects. So that when your muscular sense impresses on you the notion that you are exerting force, as in pushing or pulling, you ought to be very cautious in forming a judgment as to what is really going on ; and you ought to demand much further evidence before admitting the objective reality of force.

Until all physical science is reduced to the deduction of the innumerable mathematical consequences of a few known and simple laws, it will be impossible to altogether avoid some confusion and repetition, whatever be the arrangement of its various parts which we adopt in bringing them before a beginner. But when we confine ourselves to one definite branch of the subject, all of whose fundamental laws can be distinctly formulated, there need be no such confusion. Here, in fact, the mathematician (who, be it most carefully observed, does not necessarily deal with algebraic symbols) has it all in his own hands. He is the skilled artificer with his plan and his trowel, and the hodmen have handed up to him all the requisite bricks and mortar.

'That which is properly called Physical Science is the knowledge of relations between natural phenomena and their physical antecedents, as necessary sequences of cause and effect ; these relations being investigated by the aid of mathematics—that is, by a method in which processes of reasoning, on all questions that can be brought under the categories of *quantity* and of *space-conditions*, are rendered perfectly exact, and simplified, and made capable of general application to a degree almost inconceivable by the uninitiated, through the use of conventional symbols. There is no admission for any but a mathematician into this school of philosophy. But there is a lower department of natural science, most valuable as a precursor and auxiliary, which we may call scientific phenomenology; the office of which is to observe and classify phenomena, and by induction to infer the laws that govern them. As, however, it is unable to determine these laws to be necessary results of the action of physical forces, they remain merely empirical until the higher science interprets them. But the inferior and auxiliary science has of late assumed a position to which it is by no means entitled. It gives itself airs, as if it were the mistress instead of the handmaid, and often conceals its own incapacity and want of scientific purity by high-sounding language as to the mysteries of nature. It may even complain of true science, the knowledge of causes, as merely mechanical. It will endue matter with mysterious qualities and occult powers, and imagines that it discerns in the physical atom, "the promise and potency of all terrestrial life."[1]

Whether there is such a *thing* as Force or not, I shall consider presently. But in the meanwhile there can be no doubt that it is a convenient term, provided it be employed in one definite sense, and one only. Let us then first see how it is to be correctly used. Here we cannot but consult Newton. The sense in which he uses the term *Vis Impressa*, the Latin equi-

[1] *Church Quarterly*, April 1876.

valent of the scientific word 'Force,' and therefore the sense in which we must continue to use it if we desire to avoid intellectual confusion, will appear clearly from a brief consideration of his simple statement of the Laws of Motion.

The first of these Laws is:—*Every body continues in its state of rest or of uniform motion in a straight line, except in so far as it is compelled by forces to change that state.*

In other words, any change whether in the *direction* or in the *rate* of motion of a body is attributed to *Force*. Thus a stone let fall moves quicker and quicker, and we say that a force (viz. the weight of the stone or the earth's attraction for it) is continually acting so as to increase the *rate* of the motion. If the stone be thrown upwards, the *rate* of its motion continually diminishes, and we say that the same Force (the stone's weight) is continually acting so as to produce this diminution of speed. So far, none of you probably feels the least difficulty. But we have got only half of the information on this point which Newton's First Law affords. You see the moon revolving about the earth, and the earth and other planets revolving about the sun—approximately at least in circles—why is this? Their *directions* of motion are constantly changing—in fact, a curved line is merely a line whose direction changes from point to point, while a straight line is one whose direction does not change—but to produce in a body this change of direction of motion Force is required just as much as to produce change of speed. That is supplied by the gravitation attraction of the central body of the system. The old notion was that a centripetal Force was required to balance the so-called centrifugal Force, it being imagined that a body moving in a circle had a tendency to fly outwards from the centre! Newton's simple Law exposes fully the absurdity of this. If a body is to be made to move in a curved line instead of its natural straight path, you must apply Force to compel it to do so; certainly not to prevent it from flying outwards from the centre about which it is for the moment revolving. In fact, the *Vis Inertiae*, now called Inertia (*not*, mark you, The Force of Inertia), implies, not revolutionary activity, but dogged perseverance; and just as you must apply force *in* the direction of motion to change the *rate* of motion, so must you apply force *perpendicular* to the direction of motion to change that *direction.*

Newton's Second Law is now required: *Change of motion is proportional to the force, and takes place in the direction of the straight line in which the force acts.*

Mark here most carefully that this simple law (upon which the *Parallelogram of Forces* depends) holds for *all* kinds of force alike. There is no special law for gravitation-force, and others for electric and magnetic forces. All are defined alike, without reference to their origin.

Motion, as Newton has previously defined it, is here used as a technical scientific term for what we now call *momentum.* It is the product of the mass moving into the velocity with which it moves. 'Change of motion,' therefore, is change of momentum or the product of the mass of the moving body into its change of velocity. Now, a change of velocity is itself a velocity, as we see by the science of mere motion—kinematics, the purely mathematical science of mixed space and time.

Newton's words, however, imply more than this. Of course, the longer a given force acts the greater will be the change of momentum which it produces; so that to compare forces, which is the essence of the process of measuring them, we must give them equal times to act, or, in scientific language, we must measure a force by the *rate* at which it produces change of momentum. Rate of change of velocity is called in kinematics Acceleration. Thus the measure of a force is the product of the mass of the body moved into the acceleration which the force produces in it. This is the so-called *Vis Motrix Impressa*, the 'moving force' of the Cambridge text-books—the *Vis Acceleratrix* or so-called 'accelerating force' being really no force at all, but another name for the kinematical quantity Acceleration which I have just defined.

Unit force is thus that force which, *whatever be its source*, produces unit momentum in unit of time. If we employ British units—unit of force is that which, in one second, gives to one pound of matter a velocity of one foot per second. Here you must carefully notice that a *pound* of matter is a certain *mass* or quantity of matter. When you buy a pound of tea you buy a quantity of the matter called tea equal in *mass* to the standard pound of platinum. The idea of weight does not enter primarily into the process. In fact, the use of an ordinary balance depends upon one clause of Newton's Law of Gravitation, which tells us that in any locality whatever the weights of bodies are equal if their masses are equal. The weight of a pound of matter varies from place to place on the earth's surface; it depends on the attracting as well as the attracted body. The mass of a body is its own property. The earth's attraction for a body, or the weight of the body, is a force which produces in it in one second a velocity which (in this latitude, and at the sea-level) is about 32·2 feet per second. So that in Glasgow the weight of a pound, which we take as our standard of mass, is rather more than 32 units of force—or, what comes to the same thing, the British unit of force is about the former limit of weight of a penny letter—that of half an ounce.

Some people are in the habit of confounding force with momentum. No one having sound ideas of even elementary mathematics could be guilty of this or any similar monstrosity. He would as soon, as Hopkins used to say, measure heights in acres, or arable land in cubic miles. But to show to a non-mathematician that it is really monstrous to confound force and momentum, it suffices to change the system of units employed in measuring them, when it will be found that, if numerically equal for any one system of units, they are necessarily rendered unequal by a mere change of the unit employed for time. Now two things which are really equal to one another must necessarily be expressed by the same numerical quantity *whatever* system of units be adopted. Let us try, then, unit of force and unit of momentum, as defined by pound, foot, second, units, and see what alterations a common change of these fundamental units will make in their numerical expression.

Unit momentum is that of one pound of matter moving with a velocity of one foot per second. Unit force is that force which, acting for one second, produces in unit of mass a velocity of one foot per second. In each of these statements you may put an ounce or a ton instead of a pound, and an inch or a mile in place of a foot, and their relative value will not be altered. But suppose we take a minute instead of a second as the unit of time. One foot per second is 60 feet per minute, so this change of the time unit increases sixty-fold the nominal value of the momentum considered. But in the case of the force our statement would stand thus:—What we formerly called unit of force is that which, acting for one-sixtieth only of our new unit of time, produces in a mass of one pound sixty-fold the new unit of velocity. In other words, the number expressing the momentum is increased sixty-fold, while that representing the force is increased three thousand six hundred fold.

In fact, whatever be the system of units you employ, if you increase in any proportion the unit of time, the measure of a momentum is increased in that proportion simply, while that of a force is increased in the duplicate ratio. The two things are, therefore, of quite dissimilar nature, and cannot lawfully be equated to one another under any circumstances whatever. The mathematician expresses this distinction at once by saying that Momentum is the Time-Integral of Force, because force is the rate of change of Momentum.

But what I have already said as to the meaning of Newton's two first laws leaves absolutely no doubt as to the only definite and correct meaning of the word Force. It is obviously to be applied to any pull, push, pressure, tension, attraction or repulsion, etc., whether applied by a stick or a string, a chain or a girder, or by means of an invisible medium such as that whose existence is made certain by the phenomena of light and radiant heat, and which has been shown with great probability to be capable of explaining the phenomena of electricity and magnetism.

I have already mentioned to you that the notion of force is suggested to us by the so-called muscular sense, which gives us a peculiar feeling of pressure when we attempt to move a piece of matter. To get a notion of what it really means we must again have recourse to physical facts, instead of the uncontrolled evidence of the senses. Almost all that is required for this purpose is summed up for us in the remaining law of Motion. Before we take it up, however, let us briefly consider the position at which we have arrived.

We have seen how to get rid of two gratuitous absurdities:—the so-called Centrifugal Force, and Accelerating Force; and we must proceed to exterminate Living Force. Cormoran and Blunderbore have been disposed of, but a more dangerous giant remains. More dangerous, because he is a reality, not a phantom like the other two. Whatever force may be, there is no such thing as Centrifugal Force; and Accelerating Force is not a physical idea at all. But that which is denoted by the term Living Force, though it has absolutely no right to be called force, is something as real as matter itself. To understand its nature we must have recourse to another quotation from the *Principia*.

Newton's Third Law of Motion is to the effect that *—To every action there is always an equal and contrary reaction; or, the mutual actions of any two bodies are always equal and oppositely directed.*

This law Newton first shows to hold for ordinary pressures, tensions, attractions, impacts, etc., that is for *forces* exerted on one another by two bodies; or their time-integrals. And when he says—If any one presses a stone with his finger, his finger is pressed with an equal and opposite force by the stone, we begin to suspect that force is a mere name—a convenient abstraction—not an objective reality.

Pull one end of a long rope, the other being fixed. You can produce a practically *infinite* amount of force, for there is stress across every section throughout the whole length of the rope. Press upon a moveable piston in the side of a vessel full of fluid. You produce a practically infinite amount of force, for across every ideal section of the liquid a pressure per square inch is produced equal to that which you applied to the piston. Let go the rope, or cease to press on the piston, and all this practically infinite amount of force is gone!

The only man who, to my knowledge, ever tried to discover experimentally what might be correctly called *Conservation of Force* was Faraday. He was not satisfied with the mode of statement of Newton's law of gravitation, in which the mutual attraction between two bodies is said to VARY inversely as the square of their distance from one another. When the distance between two bodies is doubled their mutual attraction falls off to one-fourth of what it formerly was. Faraday seriously set to work to determine what became of the three-fourths which have disappeared, but all his skill was insufficient to give him any result. Faraday's insight was so profound that we cannot assert that something may not yet be discovered by such experiments; but it will assuredly not be a conservation of force.

But Newton proceeds to point out that there is another and much higher sense in which his statement is true. He says:—

'*If the action of an agent be measured by the product of its force into its velocity ; and if, similarly, the reaction of the resistance be measured by the velocities of its several parts into their several forces, whether these arise from friction, cohesion, weight, or acceleration;—action and reaction, in all combinations of machines, will be equal and opposite.*'

The actions and reactions which are here stated to be equal and opposite, are no longer simple forces, but the *products* of forces into their velocities, *i.e.* they are what are now called *Rates of doing Work;* the time-rate of increase, or the increase per second of a very tangible and real SOMETHING :—for the measurement of which rate Watt introduced the practical unit of a *horse-power*, or the rate at which an agent works when it lifts 33,000 pounds one foot high per minute against the earth's attraction.

Now, think of the difference between raising a hundredweight and endeavouring to raise a ton. With a moderate exertion you can raise the hundredweight a few feet, *and in its descent it might be employed to drive machinery, or to do some other species of work;* but tug as you please you will not be able to lift the ton, and therefore after all your exertion it will not be capable of doing any work by descending again.

Thus it appears that *force* is a mere name ; and that the *product of a force into the displacement of its point of application* has an objective existence. [Even those who are so metaphysical as not to see that the product of a *mere name* into a *displacement* can have objective existence, may perhaps see that the quotient of a horse-power by a velocity is not likely to be more than a mere name.] In fact, modern science shows us that force is merely a convenient term employed for the present (very usefully), to shorten what would otherwise be cumbrous expressions ; but it is not to be regarded as a *thing* any more than the bank *rate of interest* (be it two, two and a half, or three per cent.) is to be looked upon as a sum of money, or than the birth-rate of a country is to be looked upon as the actual group of children born in a year. Another excellent instance is to be had from the rainfall. We say rain fell on such a day at the rate of an inch in twenty-four hours. What *can* be an inch of rain? especially when we mean a *linear* not a *cubic* inch. But there is no confusion or absurdity here. What is implied is, that, if it had gone on raining at that rate for twenty-four hours, and if the rain (like snow) remained where it fell, the ground would have been coated to the depth of an inch.

In fact, a simple mathematical operation shows us that it is precisely the same thing to say:—

The horse-power of an agent, or amount of work done by the agent in each second, is the product of the force into the average velocity of the agent,[1]
and to say:—

Force is the rate at which an agent does work per unit of length.

In the special illustration of Newton's words which I have just given, the resistance was a *weight*, that of a hundredweight or of a ton. When the resistance was overcome, work was done, and it was stored up for use in the raised mass—in a form which could be made use of at any future time.

Following a hint given by Young, we now employ the term ENERGY to signify the power of doing work, in *whatever* that power may consist. The raised mass,

---

[1] [I have slightly altered this sentence so as to meet the preposterous charge, made by some of my critics, that I spoke here of the 'horse-power done in a second'! The preternatural acumen, required to make this discovery, ought to have been carefully reserved for some altogether illegible inscription in some unknown language.]

then, we say possesses in virtue of its elevation an amount of energy precisely equal to the work spent in raising it. This dormant, or passive, form, is called *Potential* Energy. Excellent instances of potential energy are supplied by water at a high level, or with a 'head,' as it is technically called, in virtue of which it can in its descent drive machinery—by the wound-up 'weights' of a clock, which in their descent keep it going for a week—by gunpowder, the chemical affinities of whose constituents are called into play by a spark, etc., etc.

Another example of it is suggested by the word 'Cohesion' employed in Newton's statement, and which must be taken to include what are called molecular forces in general, such as, for instance, those upon which the elasticity of a solid depends.

When we draw a bow we do work, because the place of application of the force exerted has a velocity ; but the drawn bow (like the raised weight) has, in potential energy, the equivalent of the work so spent. That can in turn be expended upon the arrow ; and *what then ?*

Turn again to Newton's words, and we see that he speaks of one of the forms of resistance as arising from 'acceleration.' In fact, the arrow, by its inertia, resists being set in motion ; work has to be spent in propelling it ;—but the moving arrow has that work in store in virtue of its motion. It appears from Newton's previous statements that the measure of the rate at which work is spent in producing acceleration is *the product of the momentum into the acceleration in the direction of motion*, and the energy produced is measured by *half the product of the mass into the square of the velocity produced in it.* This active form is called

Kinetic Energy, and it is the double of this to which the term *Vis Viva*, erroneously translated *Living Force*, has been applied.

As instances of ordinary Kinetic Energy, or of mixed kinetic and potential energies, take the energy of a current of water, capable of driving an undershot wheel; of winds which also are used for driving machinery; the energy of water-waves or of sound-waves; the radiant energy which comes to us from the sun, whether it affect our nerves of touch or of sight (and therefore be called radiant heat or light) or produce chemical decomposition, as of carbonic acid and water in the leaves of plants, or of silver salts in photography (and be therefore called actinism); the energy of motion of the particles of a gas, upon which its pressure depends, etc. [When the motion is vibratory the energy is generally half potential, half kinetic.]

These explanations and definitions being premised, we can now translate Newton's words (without alteration of their meaning) into the language of modern science, as follows:—

Work done on any system of bodies (in Newton's statement, the parts of any machine) *has its equivalent in work done against friction, molecular forces, or gravity, if there be no acceleration; but if there be acceleration, part of the work is expended in overcoming the resistance to acceleration, and the additional kinetic energy developed is equivalent to the work so spent.*

But we have just seen that when work is spent against molecular forces, as in drawing a bow or winding up a spring, it is stored up as potential energy. Also it is stored up in a similar form when done against gravity, as in raising a weight.

Hence it appears that, according to Newton, whenever work is spent, it is stored up either as potential or as kinetic energy—except possibly in the case of work done against friction, about whose fate he gives us no information. Thus Newton expressly tells us that (except possibly when there is friction) *energy is indestructible*—it is changed from one form to another, and so on, but never altered in quantity. To make this beautiful statement complete, all that is requisite is to know *what becomes of work done against friction.*

Here, of course, experiment is requisite. Newton, unfortunately, seems to have forgotten that savage men had long since been in the habit of making it whenever they wished to procure fire. The patient rubbing of two dry sticks together, or (still better) the drilling of a soft piece of wood with the slightly blunted point of a hard piece, is known to all tribes of savages as a means of setting both pieces of wood on fire. Here, then, heat is undoubtedly produced, *but it is produced by the expenditure of work.* In fact, work done against friction has its equivalent in the heat produced. This Newton failed to see, and thus his grand generalisation was left, though on one point only, incomplete. The converse transformation, that of heat into work, dates back to the time of Hero at least. But the knowledge that a certain process will produce a certain result does not necessarily imply even a notion of the 'why;' and Hero as little imagined that in his æolipile heat was *converted into work*, as do savages that work can be *converted into heat.*

But whenever any such conversion or transference takes place, there is necessarily motion; and the mere rate of conversion or transference of energy per unit length of that motion is in the present state of science very conveniently called Force. No confusion can arise from using such a word in such a sense. On the contrary, there is always a gain in clearness, when compactness can lawfully be introduced.

Rumford and Davy, at the very end of last century, by totally different experimental processes, showed conclusively that the materiality of heat could not be maintained; and thus gave the means of completing Newton's statement, which, still further extended and generalised, rather more than thirty years ago, by the magnificent experimental work of Colding and Joule, now stands as one massive pillar of the fast-rising temple of Science—known as the law of the *Conservation of Energy*.

The conception of kinetic energy is a very simple one, at least when visible motion alone is involved. And from motion of visible masses to those motions of the particles of bodies whose energy we call Heat, is by no means a very difficult mental transition. Mark, however, that heat is not the mere motions, but the energy of these motions—a very different thing, for heat and kinetic energy in general are no more '*modes of motion*' than potential energy of every kind (including that of unfired gunpowder) is a '*mode of rest!*' In fact, a '*mode of motion*' is, if the word motion be used in its ordinary sense, purely kinematical, not physical:—and, if motion be used in Newton's sense, it refers to momentum, not to energy.

The conception of potential energy, however, is not by any means so easy or direct. In fact, the apparently direct testimony of our muscular sense to the existence of force, makes it at first much easier for us to conceive of force than of potential energy. *Why* two masses of matter possess potential energy when separated—in virtue of which they are conveniently said to attract one another—is still one of the most obscure problems in physics. I have not now time to enter on a discussion of the very ingenious idea of the ultramundane Corpuscles, the outcome of the lifework of Le Sage, and the only even apparently hopeful attempt which has yet been made to explain the mechanism of Gravitation. The most singular thing about it is that, if it be true, it will probably lead us to regard all kinds of energy as ultimately Kinetic.

And a curious quasi-metaphysical argument may be raised on this point, of which I can give only the barest outline. The mutual convertibility of Kinetic and Potential energy shows that relations of equality (though not necessarily of identity) can exist between the two; and thus that their proper expressions involve the same fundamental units and in the same way. Thus, as we have already seen that kinetic energy involves the unit of mass and the square of the linear unit directly, together with the square of the time unit inversely; the same must be the case with potential energy. And it seems very singular that potential energy should thus essentially involve the unit of time if it do not ultimately depend in some way on energy of motion.